TIGERS

TIGERS

Michael W. Richards

Hashim Tyabji

NEW
HOLLAND

First published in 2008 by New Holland Publishers
London • Cape Town • Sydney • Auckland

www.newhollandpublishers.com

Garfield House, 86–88 Edgware Road, London W2 2EA, United Kingdom
80 McKenzie Street, Cape Town, 8001, South Africa
Unit 1, 66 Gibbes Street, Chatswood, NSW 2067, Australia
218 Lake Road, Northcote, Auckland, New Zealand

10 9 8 7 6 5 4 3 2 1

ISBN 978 1 84773 111 1

Although the publishers have made every effort to ensure that information contained in this book was
meticulously researched and correct at the time of going to press, they accept no responsibility for any
inaccuracies, loss, injury or inconvenience sustained by any person using this book as reference.

Publishing Director: Rosemary Wilkinson
Commissioning Editor: Simon Papps
Project co-ordinator: James Parry
Editor: Marianne Taylor
Designer: Namrita Price-Goodfellow (D & N Publishing)
Production: Melanie Dowland

Reproduction by Pica Digital PTE Ltd, Singapore
Printed and bound by Tien Wah Press, Singapore

Special thanks to Philip Dalton

All images by John Downer Productions except the following:
Pages 20, 21, 31 (both images), 94, 120 and 142–143 © NHPA; pages 38, 60, 61, 71 and 109 ©
FLPA; pages 132–133 © Nature PL; page 70 © Satyendra K Tiwari.

DEDICATION

In memory of Mohan

CONTENTS

A Cameraman's Diary

Special Features

Mike and Geoff with their elephants and the mahouts

FOREWORD

After many years of studying tigers at Nagarahole, in my home state of Karnataka, I visited Pench National Park in 1994 to take up the first-ever camera-trap study of tiger numbers in Madhya Pradesh state.

I was astonished by the ecological productivity of Pench forests in comparison to the better-known Kanha reserve nearby. Much of the productivity that supports high densities of tiger's prey, like chital and barasingha, seen in Kanha is concentrated around the meadows, which are remnants of abandoned fields. The *Sal* forests that surround these meadows are, however, lower in forage productivity and consequently support lower densities of ungulates. On the other hand, the teak and arjuna forests of Pench provide abundant forage all over. Besides, the moisture of the Pench reservoir basin creates dry season grasslands in which ungulates thrive.

My study in Pench was of a short duration, a few months in early 1995, but its results were revealing. Estimates of prey density showed that Pench was effectively a supermarket for big predators. It supported an astonishing population density of 64 ungulates per square kilometre. Major prey like chital and sambar were particularly abundant, packed at densities of 51 and 10 animals per square kilometre respectively. I guessed that at such high prey densities Pench should teem with tigers: at densities of 10-12 tigers per 100 square kilometres as in the forests of Kanha and Nagarahole.

However, from camera trap studies I had developed found that tiger density was less than half of what I could predict. The camera trap photographs showed four adult females which appeared to be residents holding relatively small territories, as expected with such high prey densities. The 'missing tigers', however, were the big males and transient floaters. The reason: Pench was under siege from organized poachers who killed off tigers that forayed out of its small confines. Forest and police officials were hot on the heels of these poachers when I left Pench.

This excellent book, with its evocative text by Michael W. Richards and Hashim Tyabji coupled with stunning images, brought back to me vivid memories of Pench. Michael and his techno-wizard colleague Geoff Bell worked in Pench during 2004-2007 to film *Tiger – Spy in the Jungle* for the BBC. This book is a narrative by Michael about their innovative natural history film-making, interwoven with informative text on tiger ecology and conservation by naturalist Hashim. The descriptions of the innovations employed by Michael and Geoff for filming tigers, like their 'Bouldercam', 'Trunkcam' and 'Tuskcam' – remotely controlled smart cameras – cleverly concealed in containers that looked like a boulder or were mounted on a riding elephant's trunk or tusk respectively, are fascinating. The resulting non-intrusive images show wild tigers, their prey and habitats of Pench in their full undisturbed glory. *Tigers* indicates that, in spite of doomsday prophesies we hear constantly, tigers appear to be better off in Pench now than they were a decade ago.

I hope tigers will continue to flourish in Pench, and the informative text and beautiful pictures in *Tigers* will contribute towards this desirable outcome.

K. ULLAS KARANTH, Ph.D.
Senior Conservation Scientist
Wildlife Conservation Society

R U S S

Mediterranean Sea

Black Sea

TURKEY

GEORGIA

Caspian Sea

K A Z A K H S T A N

LEBANON

SYRIA

AZERBAIJAN

UZBEKISTAN

Syr Dar'ya

ISRAEL

JORDAN

IRAQ

TURKMENISTAN

Amu Dar'ya

KYRGYZSTAN

IRAN

TAJIKISTAN

KUWAIT

AFGHANISTAN

SAUDI
ARABIA

QATAR

UNITED ARAB
EMIRATES

PAKISTAN

Indus

NEPAL

BHUTAN

Ganges

Arabian Sea

BANGLADES

YEMEN

OMAN

I N D I A

Bay of Bengal

SRI
LANKA

Corbett

N E P A L

Chitwan

Sariska

Ranthambore

I N D I A

Bandhavgarh

Kanha

Pench

I N D I A N   O C E A N

Tiger habitat, 1900

Current or potential tiger habitat

0

0   1000

MONGOLIA

CHINA

Amur

Huang He

Chang Jiang

NORTH
KOREA

SOUTH
KOREA

Sea of Japan

JAPAN

East China Sea

PACIFIC OCEAN

TAIWAN

Irrawaddy

MYANMAR

VIETNAM

LAOS

Mekong

THAILAND

CAMBODIA

South China Sea

PHILIPPINES

PAPUA
NEW
GUINEA

BRUNEI

MALAYSIA

SINGAPORE

Sumatra

Borneo

INDONESIA

1000 miles

2000 kms

Java

Bali

AUSTRALIA

INTRODUCTION

Tigers... *Spy in the Jungle* is the latest film in the highly successful *Spy in the...* series that I have filmed for John Downer Productions over the past eight years.

The previous programmes in the series featured Lions, African Elephants, bears and Blue Wildebeest, but Bengal Tigers presented perhaps our greatest challenge; we had to apply the photographic techniques we had developed over the preceding years to tell a story about tigers in the more photographically demanding jungles of India.

There have been many films on tigers over the years – not surprising for such a charismatic animal. But it was more than a little daunting starting our own film project, as we knew from the start that it had to offer something new – a fresh perspective and a new insight, something that would not only do the tigers justice but would also help this beleaguered species in some way. With a subject that lives in dense jungle, survives by camouflage and depends on the element of surprise for attack, it was never going to be easy.

We had great confidence in the array of cameras and techniques that had been so successful in the other *Spy* films, but we were determined this time to film our subject on the move and, ideally, at the tiger's level. Previous film-makers working with tigers had relied on using a camera with an extremely heavy long lens, attached to a huge unwieldy tripod that the film-maker would erect beside the elephant on which he or she was mounted. Often the action would be over by the time the tripod had been put in place. We needed something far more mobile.

Fortunately, technology is a wonderful thing and the pace of change is quicker than we could have imagined when we started making wildlife films – now we had inventive and intriguing new options for filming the tigers. At the beginning, we only had a basic plan, lots of questions and no possible way of knowing how it was all going to work. With the help of a host of people we settled on Pench National Park, in Madhya Pradesh, to work out just how we could get close to our elusive subject.

Pench was chosen for several key reasons. First, it has a population of tigers – a pretty important requirement! Second, it has trained Asian Elephants and their mahouts – vital if our embryonic filming plans had a chance of working. Third, it is a relatively new addition to India's Project Tiger scheme to establish protected National Parks for tigers, therefore little known outside India compared to parks like Corbett and Ranthambhore so it had very few visitors. Fourth, the park authorities were happy, indeed keen to have us, recognising that our film would help promote the park.

Apart from the tigers, the elephants were perhaps the most important requirement. Elephants are able to negotiate dense jungle, they can effortlessly carry a film crew and kit, but most importantly they can get close to tigers without disturbing them. Tigers accept elephants as part of their world – there is a mutual respect between the two species that we could use to our advantage to get close to our subject. The mahouts who look after and ride the elephants also have an incredible knowledge of the tiger's behaviour and whereabouts.

If our plan was to work, the elephants had to do something even more important. They would have to carry the cameras that would capture the tigers on film – they would effectively become elephant cameramen!

On previous productions we had used a device called 'Bouldercam'. This was a mobile remote-controlled camera disguised as a rock, which could be driven towards its subject and capture up-close and personal images. Bouldercam had been used very successfully on Lions, so tigers were a natural progression – but would the elephants carry it?

We attached a carrying handle to Bouldercam and auditioned the Pench elephants. Their first attempts weren't promising. They would pick it up okay, but when they came to put it down they would just throw it to the ground. Geoff Bell, who had built the device, looked on with increasing alarm as his 'baby' began to shed vital parts. Our plans were soon in almost as many pieces as Bouldercam. But then something remarkable happened. 'Mohan', an old bull elephant, picked up the device as gently as a mother would a baby and carried it through the forest, carefully avoiding any obstruction. Then, just as gently, he placed it down again. Suddenly filming in this way seemed to be a real possibility.

Mohan was 50 years old. He had a lovely temperament and was to prove a gift to the project. There are only four working elephants in Pench – the numbers are kept low to allow the forest to regenerate after decades of logging. Elephants eat and destroy a lot of trees, especially young saplings, and since the park was made a reserve in 1975, the aim has been to allow the natural vegetation to grow back.

The park elephants have a history of working in the forests, so looking for tigers is natural for them and their mahouts. They take tourists to see tigers and researchers to study them, and they are also used for a few timber-related tasks around their camp in the park. During the monsoon, a period of three months of heavy rain when the park is closed, the mahouts patrol the borders on their elephants to deter poachers.

Bouldercam is a very sophisticated piece of engineering. Its inventor, Geoff Bell, is with us all the time when we are filming. Geoff has been a key element in the success of the *Spy* films with his innovative approach to problem-solving and his ability to create wonderful pieces of engineering that have enabled the camera devices to evolve with the requirements of the films.

Filming tigers presented a new set of challenges, which he rose to with characteristic enthusiasm. His boffin-like solutions, even if they initially looked quite simple, were totally baffling once he explained how they work – even now I am still not sure what the thrusters do, but as long as they work, then I'm happy. But with all this complexity and engineering, Geoff was naturally nervous of trusting an elephant to carry his sophisticated devices.

Bouldercam is a motorized device on wheels that carries a large High Definition camera. It works by radio control over rough ground and can move in any direction. Inside, the camera is also controlled by radio. It can be turned on and off, zoomed, focused and set to record, all at the flip of a switch. It can also level its horizon and be left in standby all day. It sends a microwave image directly from the camera viewfinder to a screen, which is attached to the *howdah* on the elephant's back, giving us true flexibility.

The elephants first carried Bouldercam as it was. Later, Bouldercam was superseded by a new tailor-made invention called Trunkcam. Trunkcam is carried by the trunk and looks like a huge tree trunk. This proved an easier device for an elephant to carry as it could use its tusks to help support it – otherwise Trunkcam worked in the same way as Bouldercam.

As the filming progressed, we developed another key device, which Geoffrey designed to be carried on the tusk of an elephant.

Tuskcam is a stabilized camera platform, and as revolutionary a device as Bouldercam was in its early days in Africa. Mohan accepted Tuskcam very quickly; he hardly noticed its weight and was able to carry it all morning without any problem. Not only that, but he was able to carry Bouldercam at the same time, making him a true cinematographer. In time we managed to get Tuskcam onto the tusk of another, much more temperamental, male elephant called 'Jung Bhadhur', as all he had to do was to keep it on his tusk.

It was a real breakthrough to have two elephants that could carry cameras and were able to track and film the tigers as they were found. I could operate the cameras from the *howdah* I was travelling in or from that on another elephant close by. This worked just as easily with the elephant tracking the tigers on the move. It was really exciting to enter the world of the tiger at this unprecedented level of intimacy.

Originally we only had Bouldercam to use. It was a real thrill when we went out with it for the first time to meet a tiger. Would the tiger accept the camera device? The only way to find out was to try it and see. Our experience with lions in Africa was that they would usually stare at the camera, watch it move for a bit, then walk up to it to investigate it more closely, sometimes patting it lightly with their paw. Once they realized it wasn't edible or dangerous they would soon lose interest and treat it like any other inanimate object. We thought there was a very good chance that the tigers, too, would be curious at first and then quickly accept the cameras.

This is more or less exactly what happened. The tigers just needed to see it operating to learn that it wasn't a threat, and then they too would ignore it. This was perfect for us, as we were then in a position to film freely, knowing that we weren't interfering with the tigers' lives or influencing their behaviour. The cameras really were spies.

We came to India many times over the two-and-a-half years it took us to make the film. We used the same camp as a base so we could leave the equipment and return for a break and work on other projects concurrently. John Downer himself came out to join Geoff and myself whenever he could. These are the notes of our day-to-day experiences while making the film.

Michael W. Richards

Pench National Park

Our arrival yesterday afternoon has certainly set the scene. Just travelling to the park answered many of our initial questions. First impressions were, as they so often are, that the task ahead will be difficult.

It is always very challenging to start a new film, as everything is new – the subject, the expectations, the setting and the people we'll be working with. There is so much to take in and arrange. I also have a whole suitcase of film that I know I'm supposed to fill with amazing images of tigers by the time we leave. Right now, I cannot see where any of these images are going to come from.

It's daunting and exciting at the same time. Tigers are such high-profile animals that we're desperate not to waste this opportunity. We feel the need to make something really special that does the tigers justice by gaining a new insight on their lives.

Today was our first day out and we were keen to establish a routine of early starts, to be in the park as soon as we are allowed. Our permit said sunrise – later than I would have liked, but whatever has been written on our permits has to stand. It was a performance to get the gates unlocked by the guards and we were informed that we have to take a park guard with us at all times. It had been a cold night and was still surprisingly cold in the early morning, especially standing in the back of our open-topped vehicle. I really needed my warm clothes, I'm glad I brought them.

Despite the cold, winter is a marvellous time of the year. It was very still and clear with lovely light, and the calls of the animals in the forest brought a great sense of excitement and anticipation.

We planned to film what we could from the vehicle and wait to hear from the mahouts, via a handheld radio, when they had found a tiger. We would then drive and meet them, get on the elephants and go and film our subject. Simple... but today it didn't happen, because no tigers were seen.

The whole tradition of looking at tigers from elephant-back goes back centuries but our ideas were anything but traditional. We have given a lot of thought as to how we would use our devices to make the film but we face so many uncertainties. We have ambitious plans and some tricks up our sleeve to try out with the elephants but it's obviously going to take a while to get the elephants involved with our ideas.

With no tigers to film, today we filmed mostly monkeys, Grey Langurs in fact. They are truly good-looking monkeys and in Pench there are plenty of them. There are also plenty of Spotted Deer, usually known as Chital, and I grabbed a few distant shots of them. It felt good to make a start of some sort.

We also took some time to drive around and get to know the layout of the park a little. It's certainly a large area – it will take a while to learn where the boundaries are, let alone the most likely places to find tigers.

The park is an old teak plantation that has been given the chance to regenerate. The river Pench runs through the middle and gives the park its name. There is a huge lake in the south, created by a hydroelectric dam on the river.

Park regulations prevent us driving off-road. It's too rocky anyway but, after the freedom of East Africa, this feels like quite a restriction. We shall have to find animals near the road. Jungle always gives limited visibility, that's just part of the challenge, but long grass adds to our problems. It reminds me of my last trip to India when, after the monsoon, I couldn't find the young peafowl that were hiding in similar grass. The tigers must love it though.

John had been doing a recce of the park in a separate vehicle. We met to compare notes and enjoy a magnificent lunch, provided with all the plates and dishes. It looked like a scene from the old colonial days. I can't see this luxury lasting!

The elephants only work in the morning, as they need time to eat, bathe and rest from wearing the *howdahs*. We used the afternoon to see how the elephants might carry the cameras.

Bouldercam had arrived from Africa meanwhile, already looking battle-weary from its campaigns with Lions.

The first hurdle was to see what the elephants thought of this strange moving contraption. We arranged to see Keyshu, the head of the mahouts, and went to the elephant camp in the middle of the park to meet him with his elephant, Mohan, a huge gentle giant.

First we had to show the elephant the device, and then let him see him how it worked. He showed little interest until Geoff made it move. As soon as Bouldercam motored towards him, Mohan backed away, snorting through his trunk, an anxious moment but fortunately short-lived. He was more surprised than frightened, and once he had seen what it was, he accepted it. Then came the interesting bit; he had to try and lift the camera without breaking it. We had adapted a metal ring and a chain to fix onto the top of Bouldercam, through which Mohan could put his trunk quite comfortably.

Using one of his mysterious commands, Keyshu was able to get Mohan to lift the camera from the back of the car. No-one dared breathe for every second that the camera was suspended in the air. If Mohan was unhappy, over £100,000 of equipment – the result of countless hours of painstaking work by Geoff – could be smashed against the ground. Bouldercam hovered in the air for what seemed an eternity, and then gradually and carefully Mohan lowered the device to the ground. Simultaneously, everyone breathed again and then we all let out a huge whoop of triumph – even the mahouts joined in the celebrations, although at this point they must have been unsure what the device was actually for.

Having done it once, Mohan repeated his feat with ease, this time even placing it in the back of the car. Our confidence riding high, we got him to do it a few times just to make sure.

This was a great start. So much hinges on an elephant being able to carry the camera into the forest. The next step will be to see how far he feels comfortable carrying what, to us, is a very heavy piece of kit.

The day cools down as soon as the sun sets. It's almost cold now, which is nice at the end of the day, and a hot shower is very welcome. The lodge is clean and spacious, a very good set-up so close to the park. There was even a hot waterbottle waiting in my bed when I returned from dinner.

A sign on the main road near Pench National Park, and the first indication that we were getting nearer to tigers.

Tuesday November 30, 2004

Our second day out was tinged with disappointment, as we still didn't manage to see a tiger, but we did see tiger pugmarks (footprints) on the sandy road. An excited guard had spotted them from his bicycle, and very kindly came to tell us about them.

It seems the guards here happily cycle round the park, not perceiving the tigers as a threat. The bad press that portrays tigers as ferocious man-eaters is often overplayed, but even I was a bit surprised by the nonchalant attitude of the guards.

The imprint from a tiger's foot in the soft ground holds a lot of information – if you know how to read it, that is. A pugmark deteriorates over time and its condition gives a clue as to when the tiger might have passed. The tiger's sex and size and the direction it's travelling are all there, written in the sand. But sometimes the mahouts see only a partial print in something like gravel, which is often invisible to me even when it's pointed out. To them it still yields vital information.

Tigers use dry riverbeds, called *nyalas*, as pathways through their territories. The forest roads are equally useful to them. Checking *nyalas* can be the easiest way to find a tiger, especially if you hear alarm calls from other animals near by. The deer call in alarm if they see a tiger, as do the langurs – hearing both together is a good sign that tigers are nearby. Sometimes the tigers themselves do the calling – a deep, mournful moan that causes the pulse to quicken.

Today the pugmarks didn't lead us to a tiger but I felt that the tigers were probably watching us; at least, I wanted to think that they were.

We were in touch, via the radios, with the mahouts all morning, but they couldn't find any other tracks. We had to resign ourselves to making a plan for tomorrow. Will it be the first day we see a tiger?

Langur monkey giving an alarm call; they give very loud, punctuated notes when a tiger or leopard is sighted.

16

Mohan carrying the Tuskcam down a forest path.

Wednesday December 1, 2004

The winter days seem too short, especially as we are so desperate to see our first tiger. But today seemed to be turning out to be just like yesterday — still no tiger sighting. Have we really chosen the right place to film? Is it all going to turn into a disaster?

Just as our hopes had all but faded, we found a pugmark placed over our tyre print on the same road as yesterday — meaning that the tiger must have passed after we left. We could even see where it had recently spray-scented a tree. Slightly nervously, as I felt sure the tiger must be watching, I got out of the car and sniffed the tree. The scent was so strong, so definite. Most of us would recognize a cat-spray smell, but I certainly wouldn't want this one in my living room. It was extremely tantalizing — we were clearly very close to a living tiger.

Geoff has now unpacked all the camera devices, and I was very keen to put out the first of the remote cameras down near the trees that the tiger has been by. These motion-sensitive cameras are triggered automatically whenever an animal passes nearby.

I set the cameras in the hope that the tiger would not leave this area. As I did so I was encouraged by the alarm calls of worried langur monkeys, very close and very loud. I found myself not wanting to wander too far from the vehicle — I definitely felt that I was being watched. We called the elephants over, but disappointingly the tiger remained elusive.

The forest is dry and still, which gives a lovely sense of peacefulness. Little seems to happen except for the occasional crash of langurs rampaging through dry leaves. The dryness will remain for months now. Day after day it's the same, clear, cool and still — cool in the mornings anyway, but by midday it's hot.

In the course of our travelling round the park we are beginning to see some of the rich variety of wildlife India has to offer, including the strange Nilgai antelope, Wild Boars, Gaur, Golden Jackals and Asiatic Wild Dogs, and even the bizarre armour-plated Indian Pangolin, a lucky fleeting glimpse. All seem to want a part in the film, as they briefly do a turn in front of our cameras.

BELOW LEFT: *The remote cameras captured the secret world of an adult and baby Wild Boar as they visit a water hole.*

BELOW RIGHT: *An Indian Pond-Heron suddenly sees itself in the camera lens — one of the many moments during filming that was totally unexpected and quite amusing.*

A langur investigates the remote camera. It was a wonderful moment when we first saw this result; the monkey was clearly fascinated by the lens.

What is a Tiger?

Tigers are without question one of the most beautiful of all the world's creatures. Their unique coloration and impeccable line and form combine to create an animal of striking drama. Yet for all its burnished beauty, the tiger is most remarkable as a superbly adapted hunter, a product of the evolutionary cat-and-mouse game between hunter and hunted. It originated in the forests of eastern Asia, where the process of natural selection created a giant cat to hunt the giant cattle, deer and wild pigs that inhabited these vast tracts of woodland. In the two million years since the tiger first appeared, the climate of the Earth has pulsed between cold, dry periods known as glacials – which have seen the extension of the ice sheets southwards – and warm, wet periods usually referred to as interglacials, which witnessed the expansion of forest and dense vegetation, the preferred habitat of the cattle, deer and pigs that formed the tiger's prey. As these species radiated south, east and west to occupy their expanding habitat, so the tiger followed.

The western radiation led to the Caspian race of the tiger; the eastern and southern radiations colonized all of China, southeast Asia and India. During the glacial periods, when enormous quantities of water were locked into ice, thereby reducing sea levels and revealing land bridges to the Indonesian archipelago, the tiger followed its prey to occupy Sumatra, Java and Bali, being subsequently isolated there by rising sea levels. At the height of its range the tiger occupied an enormous swathe of territory stretching from Georgia and Armenia at the edge of Europe east to Korea and Japan (which was occupied by a small race of tiger until about 12,000 years ago); from Siberia in the north to Bali, south of the Equator; in total, an area that spanned some 100 degrees of longitude and over 70 degrees of latitude!

Thus we have an animal that is both robust and adaptable, not some delicate creature confined to a narrow ecological ledge and

The stare of a stalking tiger.

The Siberian Tiger is a heavier animal with a longer coat to cope with the cold.

thereby susceptible to every minor shift in the environment. So long as adequate food, water and cover are available, it will thrive in conditions ranging from the frigid taiga of the north to lush tropical rainforests; from mangrove swamps to dry deciduous jungles or the tall dense grasslands of the Caspian and sub-Himalayas. But for all its adaptability it remains an animal of dense cover. Natural selection has honed a powerful stalk-and-ambush predator designed to hunt prey weighing up to 1,000 kg (2,200 lb). A light, supple but strong skeleton supports enormous muscles, especially the short, powerful forelegs, which are aided by formidable retractile claws — needed to hold and handle prey that is often three times a tiger's own weight. The short-muzzled head gives the jaw-closing muscle added leverage, allowing the large canines to be driven with great force. Behind the canines is a gap which allows them to sink in deeply, enhancing the effectiveness of the killing bite. Any hunter — but particularly one that needs to judge distance correctly in order to time the final charge — needs both binocular vision and reasonable acuity. The tiger has both of these, as well as a many-layered structure behind the retina called the *tapetum lucidum*, which helps reflect light back into the retina and thus aids vision in low light conditions. Finally, this entire vital package comes enveloped in a striking pelage that efficiently allows the tiger to conceal its great bulk as it stalks in close or waits in ambush for its prey.

One other factor limiting the tiger to its forest habitat is its intolerance of excessive direct heat. This is an animal that evolved in the cold lands of the north and deals far better with cold than heat. Hence its need of water and deep cover as a key element for its well-being in the warmer areas of its range. Reflecting the range of habitats they occupy, tigers were traditionally divided into eight races — a number deeply embedded into tiger lore. However, modern research, especially by taxonomy expert Andrew Kitchener of the Royal Museum of Scotland and colleagues, suggests that the earlier racial classification was based on too few scientific samples (only 11) and that 'racial' characteristics are more graduated than distinct. This view holds that there are considerable physical variations within discrete tiger populations, and not just between geographically distant populations as traditionally recognized. Accordingly Kitchener proposes only three races of tigers — the Western or Caspian (now completely extinct), the Eastern and south Asian mainland race, and the island race from Sumatra, Java and Bali (now extinct in the latter two islands). The debate continues, with a more recent study by Shu-Jin Luo and colleagues, based on molecular genetic analysis, identifying six living sub-species of tigers.

The first tiger sighting of the trip was a moment of high excitement; here it is walking through an area of forest that is characteristic of Pench.

An exciting day! Finally we have found our tiger and the whole experience is transformed. As each day passed the mounting anticipation of seeing a tiger had become unbearable, so the excitement and relief at seeing one at last is truly fantastic. It appeared so suddenly in front

of us, unconcernedly padding silently over the crisp dry leaves… a wild tiger and somehow I was filming it.

The elephants found it and the mahouts called us over to the nearest road where we climbed onto the *howdahs* of two more elephants, and set off with Mohan carrying Bouldercam in his trunk.

In Bouldercam Geoff has created a system that can go where no other camera has before. It was wonderful to watch Mohan as he managed to carry the camera several hundred metres through the thick forest undergrowth without hitting anything – the camera seemed to glide through the forest. He carried it as if his life depended on it, guided along by Keyshu in a slow methodical plod.

Our tiger was a two-year-old male but, to our untutored eye, looked like an adult. The mahouts know him of course, and told us he had been with his family until recently. Once Mohan put Bouldercam down we cautiously moved it towards the tiger to attempt to begin filming him. It was an important test for Bouldercam. Would the tiger be scared, or attack the camera as possible prey? Fortunately he showed only a casual interest, soon realising that it wasn't a threat. We could now start filming in earnest.

It seemed clear that any tiger that accepted an elephant would also accept Bouldercam – a huge boost for our spirits. Our excitement increased even more when the young tiger suddenly leapt into the undergrowth to chase and kill a deer fawn. The image of him bounding, all four legs at full stretch and tail high above the grasses, was electrifying. The sheer power and agility with which he moved left me in no doubt as to how easily tigers can catch their prey. I made a rather shaky recording of the event but it was enough to give us a leap of confidence. Here was a situation that could easily repeat itself and gave us one of our building blocks for telling the tiger's story – hunting is such an important part of their lives and so difficult to see. Witnessing such excitement fired up our confidence that, yes, there is a great film to be made about tigers here in Pench.

This tiger was clearly staring at deer in the distance.

Saturday December 4, 2004

Another great day! A tiger was found early today and in a different area. It was hidden in lantana bushes. Initially I thought this would be a problem as the shrub grows so densely it is virtually impossible to see into. However, Bouldercam on the ground sees the world as the tiger sees it. It looks through and under the bushes and even shows the obvious routes in and out that the tigers take. It transforms everything, giving exciting images of the tiger's world.

Lantana (*Lantana camara*) is an alien shrubby bush introduced from South America. It is a fast-growing weed that blankets the ground, and nothing else grows under it. Tigers love it as they can hide underneath it and keep cool. For this reason the park management has decided to keep it in certain parts of the park. These areas are conveniently named Lantana No 1, No 2 and so on, up to about No 7. They are big spaces and within them tigers are difficult to locate.

Bouldercam makes an appearance. The tiger quickly accepted the camera and thereafter just ignored it.

This morning it was great to be able to get onto the elephants earlier, even though each day the park seems to get bigger. Nevertheless, we are getting much more of a feel as to how well it's all going to work, and we are starting to get a sense of how we are going to tell the story. We know that Mohan can carry the camera to the tiger. He is quick to put it down and, after the tiger leaves; he quickly picks it up again and walks with it behind the tiger.

Much of the tiger's day is spent sleeping. The winter sun is still quite hot during the middle of the day.

OPPOSITE: *The tiger walks towards the camera focusing intently on the python, which it saw very quickly even though it wasn't moving.*

LEFT: *Indian Pythons have cryptic markings but are still quite visible when they move away from dead leaves; this one was basking in the early morning sunshine.*

BELOW: *Just like the domestic cats, the tigress is always inquisitive to play with new toys, even if they can bite!*

Extraordinarily, the tiger doesn't seem to mind any of this at all, not even the constant radio chatter from the mahouts. They are far too noisy to my ears and, as I'm used to working in silence, this is a factor I am finding hard to come to terms with, but if the tiger can ignore it then I suppose I can.

While we were watching, the tiger discovered a large Indian Python sunning itself while digesting a meal. The tiger was slightly nervous of the snake and clearly knew what it was. After a little consideration it reached out a paw and patted the python. The snake was unimpressed and decided to strike out at the tiger. The dramatic moment happened very quickly – the tiger backed off, rather surprised. The encounter was quite unexpected and provided a great little sequence that will definitely be in the film.

To complete a perfect day, in the afternoon we saw the tiger from the car and got a scenic shot of it by the open river. Then suddenly it charged at a Chital in the long grass, and caught it. I was already filming it when it rushed through the grass, so I got my first full-frame tiger running, filmed from the car as it rushed by. This is so encouraging, as filming from the vehicle will be an excellent back-up when we cannot be on the elephants.

Sunday December 5, 2004

Today we found a tiger early on, but we quickly realized that this was a different animal, bigger, thicker round the face and heavier. The mahouts explained that this was the father of the male we were filming yesterday. His full belly suggested he had eaten the rest of the deer that his son had caught yesterday evening.

It soon became obvious that this tiger had no canine teeth, and so must struggle to catch food on his own. I hesitate to call him 'Gums'... but what other name could I give him?

Gums went off in search of a Chital fawn, which was calling in the tall dry grasses. Long grass is a photographer's nightmare but I got some lovely shots as he bounded through it. So exciting, but it all happens so fast, although for once we were in the right place at the right time.

These events are always tinged with a slight sense that the views could have been so much better, but witnessing hunting on two consecutive days is incredible, and beyond what we had hoped for. These moments will be the making of the film as hunting is such an important aspect of any story on tigers.

The tiger missed the fawn, and I wondered how long he could survive with no canines. Maybe he relies on his son's hunting. Mohan followed him with Bouldercam until the tiger reached the edge of a lake. He entered, gingerly as cats do with water, and swam to the other side. He was seeking the shade of the lantana and was soon fast asleep in the shadow of their arching leaves – a beautiful image which I managed to film for once!

We had eight hours on elephant-back today and I was pleased to still be able to walk again after so long. Dismounting is a bit like stepping off a boat – the sense of motion remains, a long time after returning to dry land.

We collected the remote cameras today. They had been triggered but the results were disappointing – no tiger photos. These cameras detect movement but can't discriminate between whatever moves into their field of view; even large insects could set them off. We have come to expect failures from them in the past but are still learning how to place them so we stand a better chance of success. I immediately reset one of them by one of the waterholes. As the dry season progresses these waterholes become vital to so many animals, so there should be a good chance that a tiger will make an appearance.

BELOW: *Tiger stalking Chital, or Spotted Deer. The tiger will only move when the deer looks away.*

BOTTOM: *The tiger leaps at the deer when it is just in range. The leaping tiger covers a huge amount of ground with each bound and a successful hunt is achieved when the deer makes a mistake.*

Shivpujan climbs onto Sarwasti. The mahouts use the front leg of the elephant to climb up. The elephant helps by raising its leg like a hydraulic lift, enabling the mahout to get onto the howdah with ease.

29

TERRITORY — *the basic social construct*

Tigers are essentially solitary. This is their defining social characteristic, evolved probably as a response to their closed forest environment, which favours the individual hunter over the group. Yet these solitary habits are not quite as deep-seated as one may think, for the tiger inhabits a complex social system defined by highly developed territorial behaviour practised by sexually mature adult individuals of both sexes, interacting constantly — if not always directly — with others of their kind.

Territory is primarily about the individual having assured access to valuable resources: food, water, shelter and breeding partners. These are vital elements for a tiger's well-being, and gaining a toehold on the territorial ladder is the challenging initiation to adulthood, marking an individual as fit to breed. The territorial imperatives for males and females vary. For the latter, the imperative is to possess an area with adequate food, water and shelter to support themselves and their cubs during lean times. For a male tiger, the overriding concern is to gain control over as large an area as possible, encompassing the territories of one or more females so as to propagate his own genes most effectively. That impulse is tempered by the need to create a stable and safe haven for the cubs, for the success of young tigers is dependent not just on the abilities of the mother as a provider but also of the father as a protector from competing males. The latter will try and kill all the dependent progeny of a previous incumbent to accelerate the onset of oestrus of the local tigresses to his own advantage.

Territory size is determined by the abundance of prey. The greater the prey density, the smaller the territory the female requires, and vice versa. Therefore a female tiger inhabiting the forests of the Siberian taiga with their low prey densities can roam over an area upwards of 200 sq km (77 sq miles), whereas a female from the moist tropical forests and grasslands of the Indian subcontinent may occupy a territory as small as 10–15 sq km (4–6 sq miles). Male territories are also correspondingly smaller. However, there is a limit to this territorial reduction and the secondary determinant of territory — an internal spacing mechanism — comes into play at about 10 sq km for females.

Tigers protect their territories primarily through advertisement, which takes the form of scent marking on trees and bushes and scrape markings on the ground, made with their hind feet and on which they will often defecate, urinate or leave a strong-smelling secretion from their anal glands. They will also rear up on their hind-legs and rake down on tree trunks, causing huge and rather intimidating scratchmarks, roll around on grass and vegetation and rub their cheeks against trees and shrubs. All these chemical signals carry a significant amount of information to other tigers that encounter them, and studies have shown that the frequency of these markings increases along the edge of the territory. These chemical and visual signs are backed up by a range of calls, and such 'multimedia' communication allows individuals to either avoid confrontation or, indeed, to initiate it. For it is clear that tigers, especially in productive areas where a high density of females produces a large surplus population, have to compete strongly for their territories and males in particular are often involved in physical confrontations.

However, not all tigers are territorial. There is always a floating population of sub-adults or young adults, and occasionally fully adult or even old tigers — so-called 'transients' — looking for space or an opportunity to establish themselves. These are the tigers that get forced onto the non-productive margins of good habitat, often into degraded forests which are shared with humans and livestock and where tiger territoriality becomes highly diluted.

Tigers will regularly scratch trees and rub their scent glands against the bark to advertise their presence to other tigers.

The spraying of trees and rocks is one of a tiger's key territorial markers.

Deep claw marks like these are a sign that this tree is regularly used by a tiger to mark its territory.

One interesting aspect often noticed by researchers is that tigers – especially males – will greedily expand into any adjacent area that has been vacated by, for example, a sudden death. In 1987 the dominant male in the Tala area of India's Bandhavgarh National Park, a tiger named Banka, extended his territory to the south when the former incumbent there disappeared. Banka had a most distinctive left hindfoot, making it easy to identify him from his tracks, and over the following months it was possible to follow his movements right across the 105 sq km (40 sq miles) of the entire Tala forest and into the fields of Tala village, where he was in the habit of preying upon cattle. For a few months of frantic and constant movement he managed to hold onto this huge territory before being pushed back by other younger males, and it was noticeable that he carried constant signs of combat, including a permanent limp in his right foreleg. This episode underlines how fraught the whole business of maintaining territory is, and indeed that in areas of plentiful prey, natural competition normally leads to a high turnover of tigers.

Wednesday December 8, 2004

Things are going well – it's the third day in a row of successful filming. Yesterday the mahouts found Gums and his son at daybreak and we filmed some interesting interaction between them. Today we spent what seemed like the whole morning on elephant-back, as a tiger had been found a long way from camp.

It was Gums again. He seems so used to the camera now that he hardly noticed it as it came boldly up to him. A few local tourists joined us today, a welcome colourful presence with their brightly coloured saris. They only saw the tiger for a few minutes but it was fascinating to watch their reactions – some were actually in tears from the sheer emotion of the experience. It made me feel very fortunate to be able to spend such a long time in the tiger's presence.

These tourists were particularly lucky as the tiger got up and slid into a pool of water to cool off, a lovely piece of film for us and a great campfire story for them.

The midday break was amusing today. As we tucked into our chapattis and rice (this is our daily fare now – the first day's feast is just a distant memory!) Mohan had an elephant-sized snack of his own. He simply pushed over a sizeable tree and then devoured all its leaves and the branches with them.

His lunch completely blocked the road, but one word from Keyshu and the obstruction was removed. It's easy to see why the park can only really sustain four elephants.

As we ate, the mahouts told us of another tiger which they had met this morning – a male that they see only infrequently. They feel sure this tiger – 'Charger' – will attack Bouldercam, as he has been known to charge the elephants. Sounds pretty exciting!

It was encouraging to hear the mahouts talking so excitedly and beginning to see the potential of this new camera technique. There was no chance of finding Charger today, but it was good to learn of his existence.

OPPOSITE: *Santosh gives Mohan giant chapattis to eat at the end of the day. Tuskcam is still attached to his tusk and makes no difference to his feeding ability; it was always removed at the end of each filming day.*

BELOW: *The tiger slipped off to wallow in the water. You can visibly see the sense of relief that this brings on a hot day.*

Thursday December 9, 2004

The mahouts had been out early, as they usually are now, and had checked the marsh and then the woods nearby. They heard deer alarm calls – deer only alarm consistently at a large predator so it had to be either a tiger or a leopard – and the discovery of fresh tiger pugmarks confirmed that it was the former. There was also a report from a worried guard on a bicycle, who had seen a tiger on the road. He had arrived in camp with a cut on his leg – we thought the worst but he informed us that the cut was sustained hurriedly climbing a tree! Sadly, his sighting was too far away for the elephants to go and search, but I am encouraged that the Park appears to hold more tigers.

Keyshu hasn't been feeling well for the last few days. By midday today he was still in his coat and headscarf, suffering with a fever. He has no direct access to a doctor and the nearest one is in the town which is about 20 km away on the main road. To save him using his bicycle we gave him a lift there.

It's a small town, with a few stalls, plenty of buildings of all sorts, and masses of people, of course. The surgery was easy to find, though you would never know it was such from the outside, or indeed from the inside either. Everyone from the village appeared to join us as we met the doctor – I think we attracted a certain amount of curiosity, as Westerners. The doctor thrust a thermometer under Keyshu's arm then quickly wrote out a prescription. The consultation and the pills had cost 100 rupees, just over one pound sterling, which may not seem much, but for many people this is a day's wage.

It appeared that nothing too serious was wrong with Keyshu, but he would be out of action for a few days, which meant that Mohan wouldn't be able to carry the camera.

We changed some of the remote cameras in the afternoon, as a few have been out for five days now. It is always exciting looking at these as you never know what might appear. This time there are some great images of drinking deer, and the magic of the film is strengthened by some wonderful background sound. No tigers yet but it's a great start.

An intent stare up into the trees at monkeys, which always give loud alarm calls when a predator is spotted. Tigers must find this very annoying but there is nothing they can do about it.

The female leopard that was being filmed decides to make a close investigation of the camera.

Sunday December 19, 2004

BELOW: *Unique footage of a tiger trying to climb a tree in a vain attempt to catch a leopard.*

OPPOSITE: *The tigress gave up, still looking cross from her experience with the rival big cat.*

After five frustrating days of not finding a tiger we finally have one today. It's a female, the same one we first found and filmed dozing in the midday heat nine days ago. It's quite a relief for us as this is the last day before we leave India for a couple of months. You really do have to earn the tiger sightings. It always seems so easy when we are with the tigers, but equally our task seems quite impossible when they cannot be found.

Today the tigress was walking so quickly through the forest that the elephant found it difficult to keep up with her. It amazes me that the elephants hardly ever lose a tiger once it's been found. In fact, I find it wonderful that the tigers don't attempt to lose the elephants, as I am sure they could do so easily if they wanted to.

We heard growling close by which I initially thought was another tiger, but we soon realized that the growls were coming from the top of a high tree, and belonged to a leopard. The tiger knew at once what it was and tried to climb the tree. She sprang at the trunk, throwing her paws around it as she tried to hang on, but the trunk was too thick and she had to jump away, looking cross as she stared up the trunk, before abandoning it as a lost cause. This could be the first time that such behaviour has been filmed.

Tigers are poor climbers compared to leopards, who use trees to escape from tigers. Tigers do not like leopards in their territory and will attack them given the chance. Here in Pench the leopards are managing to hold their own; the mahouts often see them.

After the tiger had given up, the leopard waited a few minutes before jumping down. It must have dropped at least 4 m before bounding away into the vegetation, seemingly unworried by the elephants standing there.

It was a remarkable sequence of behaviour and we felt privileged to witness it. I had been able to film it all with our new HD camera from the back of the elephant – it was all a bit 'grabbed', but it gave all of us confidence that we were on course to make a fantastic film here!

Our guard had dutifully been with us every day from the start. Today he asked to stop at the small temple in the elephant camp. He wanted to give thanks for the success we had enjoyed. He had a little bag of rice and coconut chips to offer inside the small concrete room. So, without his shoes, he went inside, lit the joss sticks and knelt down while we waited outside. Five minutes later he came out to offer us some of the coconut, which I felt very happy to accept.

REPRODUCTION

For many observers of tigers, among the most exciting and evocative sounds of all are those made by a courting pair serenading each other. When heard at night, against a landscape of fields and forest dimly visible in the silvery glow of the moonlight, the long roaring calls – which presage actual mating – produce a thrilling vibration in the air that seems to epitomize all that is wonderful about wilderness areas and the wildlife they support.

Such 'dueting', as it were, is the final stage of communication between courting tigers, a process that starts when the female comes into oestrus. This state is signalled chemically through the evidence she leaves in her scent marking, and which other tigers 'taste' by forcing the scent over a nasal organ called the Jacobson's or vomeronasal organ. The characteristic grimace that tigers make while doing this is known as a flehmen. Male tigers have often been observed sniffing with considerable interest at every tree sprayed by a female before making a flehmen and then beginning to call. Whether or not males always initiate the calling is unknown, however.

The oestrus cycle seems to be repeated every three weeks or so and the female is receptive for about three to five days. During this period a couple will mate up to 50 times a day, and over the entire courting and mating period males are often observed to visibly lose condition and become very bad-tempered indeed. On one occasion in India a park worker was found by his colleagues quaking high up a tree, his bicycle flung to one side – he had been chased by a male tiger normally quite placid and far too grand to take notice of a mere cyclist!

Tigers become sexually mature at about three years of age, although there are reports of some tigresses reproducing at a younger age. In the wild, the key requirement in order for a tiger to breed is the possession of a territory or home range, and although there are cases where tigresses have mated with more than one male, it appears that they will actively seek out the dominant local male. In one territory in Bandhavgarh a female tiger mated first – somewhat perfunctorily, by all accounts – with

Territorial male tigers have access to one or more breeding females.

an old resident male, who had recently been relegated to transient status by a younger rival. There was no issue, but a few weeks later she mated with the new dominant male and thereafter produced a litter. So while there will always be exceptions, the evidence suggests that females holding territory will mate with the local resident male, who will then provide vital protection from infanticidal strangers. Females with cubs will not come into oestrus or mate with any male until their cubs are about 15 to 18 months old.

The gestation period is about three and a half months and the cubs are invariably born in a secure place — a cave, rock shelter or hollowed tree root. Tigresses are notoriously fierce in the protection of their cubs, which are born blind and helpless, and they will sometimes even take on strange males that approach too closely. Humans on foot are promptly chased off. The threats to cubs are myriad: leopards, bears, wolves, wild dogs, elephants, humans, forest fires and floods all pose tremendous risks. During the initial few weeks a tigress remains close to the cubs, moving them from time to time and then starting to bring them food. After about two to three months the cubs are old enough to be brought to kills, although they will not be fully weaned until five to six months (or even longer). They will sometimes be tolerated at kills by their father, and in such circumstances the fathers have

even been observed making the call that mothers use to contact and summon their young to a kill.

An example from Bandhavgarh demonstrates that paternal tiger affection can go further than mere tolerance. Sita, a famous and much-filmed and -photographed tigress, had clearly tired of her first litter of two cubs (one male, one female) after 16 months or so and began to actively drive them off. Still untutored in the art of hunting, they were too young and inexperienced to make it on their own. They remained in the natal area, and although Sita made a few training efforts, it was clearly the father's sharing of his kills with the youngsters that helped them survive. This bond between father and offspring demonstrates the rich variety of behaviour in which tigers indulge, and perhaps reflects also variations in individual personality and character. Interestingly, the male cub maintained the bond with his father and associated peaceably with him after long absences. On one occasion the father had been tranquilized and treated for an injury. Throughout the process the son remained within 200 metres, clearly curious if not anxious, and promptly moved to be close to his father once the vets had withdrawn.

We are back again after more than three months away. Pench is a simmering wall of heat, dust and smells, and it's so, so dry and hot now winter's over. The forest has lost its leaves and there's very little water anywhere. To provide some respite for the wildlife the guards fill a few concrete basins scattered throughout the forest with water every day. Maybe a tiger will visit at night. Everything is thirsty — tiny puddles in empty dams attract swarms of bees to drink.

Mohan is on sick leave. The *howdah* ropes have rubbed sores under his belly and he needs to rest. The vet has applied a soothing cream. Hopefully we will be able to use him soon. He's such a wonderful elephant and so important to us.

We had news today that a tiger has been found near the boat camp — I investigated on the female elephant 'Sarwasti'. It was our old friend Gums, but he was obviously hot and feeling lethargic, and just rested, panting, by a rock. So, not much useful film, but at least we had a tiger on the first outing.

Jung Bhadhur carrying Tuskcam. With a little practice it became possible to attach the equipment to his tusk without any complaints from him.

With Mohan off for at least the next three days, we needed to see if any of the other three elephants might carry Bouldercam. Besides Sarwasti, who has an 18-month-old baby called 'Pench Bhadhur', there is a young female named 'Domini' and lastly the very impressive male Jung Bhadhur. This elephant has quite a reputation. Before he became a working elephant he killed 30 people while raiding their crops. He was captured and apparently is a reformed character, but his past reputation makes me nervous. He's also huge but he was the only elephant besides Mohan with tusks (female Asiatic Elephants do not grow them) so we had to see what he might be capable of.

Surprisingly, Jung Bhadhur proved very good at picking Bouldercam up when encouraged by his mahout, Satyanarayan. The problem was putting it down. Much to Geoff's consternation, he just threw it to the ground. This is what he does when they collect firewood from the forest and no matter how hard his mahout tried he couldn't get the elephant to break the habit.

We gave him something less valuable — a bucket — to practice with and turned our attention to the other elephants. Sarwasti is a slightly younger elephant, and we had high hopes she would get the idea more quickly, but her reaction was even more violent. She tossed the device like an Olympic hammer, causing damage that took hours to repair — luckily mainly dents and trouble with the electrics, nothing too serious. Clearly persuading the other elephants to carry the camera devices is going to be difficult, but I am sure we will find a solution.

Setting off in the morning with baby elephant Pench Bhadhur, who was having a first training exercise.

Today there were the first clouds in the sky, and it was muggy by the end of the day. We have heard a whisper that there might be tiger cubs somewhere in the park! We have spoken to the Deputy Director about it – he alone will say when the mahouts can visit the cubs. He agreed we can visit them, two days a week. Very exciting news.

It was another very busy day with the remotes; they are really delivering some useful material of the other animals, including lots of bees, and today some great shots of wild dogs coming down to drink at the far waterhole. There was a fresh hole just behind the camera, which has clearly been dug by a Sloth Bear for water; we could see the prints in the sand. It was close to the waterhole but dug so that the sand filtered the water for a cleaner drink – very clever as the waterhole water is beginning to look green. It's a shame that there is nothing on the camera. It happened at night and, listening to the sound on the black screen, we could hear a snuffling sound that had to be the bear.

We have also had some success with a Golden Jackal den, capturing footage of the mother and cubs. They were initially nervous of the camera, but soon got used to it and now we were getting some beautiful material.

As so much film is wasted running off at night on the remote cameras, we have asked Geoff if he can make them only work in the day, so we will have more film available. He thinks it's possible.

We found a freshly killed male chital today with no obvious cause of death. It was in a surprisingly open area and had attracted about 30 vultures. We were able to introduce Bouldercam from the vehicle and

Jung Bhadhur being bathed at the end of the day by his helper Balam, who is able to control him simply by pulling his ear. The elephants know the routine well.

slowly motored it over toward the carcass. Incredibly, the vultures were completely unfazed by it, and we got some lovely film of them squabbling. Just the same experience as we've had in East Africa. It was good to see so many vultures here, as several of the Indian species have suffered a huge population decline throughout their range – poisoned by eating cattle carcasses affected by the widespread use of Diclofenac, a drug used for pain relief to ailing livestock. Happily, once the cause of the vulture deaths was discovered, the Indian government announced that the drug's use would be phased out.

An idyllic scene of the elephants having their afternoon bath. This is one of the many waterholes created by making dams, thus ensuring a constant supply of water to the park.

This afternoon was slow. The heat creates a stifling atmosphere, the trilling of doves and calling of parakeets mark the passing time. All around is the high-pitched whine of cicadas.

The deer lie down in whatever shade they can find. The trees are without leaves or fruit. No animal wants to move. Peafowl panting somewhere shaded, run away from cars. Then the breeze comes and the elephants have their daily bath, the highlight of their day. There's no real relief for the humans though. The guard house is full of flies.

Wednesday April 13, 2005

Today we tried a new filming device, designed to be carried on an elephant's tusk, on Mohan as he is so steady. He accepted it, slid onto his tusk, as though it was a new piece of jewellery. I wondered what he really thought of it. Then I thought of the lions on which scientists place radio-collars in Africa. Apparently, it raises the social status of a lioness within the pride when she gets a collar!

Geoff has done a wonderful job creating 'Tuskcam' as it's now called. It is a round fibreglass housing for a camera that looks like a piece of wood. Inside is a self-stabilising metal disc, which the camera sits on. This has the same controls as Geoff's other creations and can be operated at the same time as rotating the housing round. Geoff has used gyros to prevent any vibration, so a really smooth image is produced on the move. When stationary the elephant can use his trunk as an extra leg for extra stability when filming is underway, a sort of elephant tripod.

OPPOSITE: *A fine view of the tiger's formidable canines as it settles down to rest during the day.*

BELOW: *The tigress walks by the water's edge. She rarely ventures far from water as the temperature is now climbing daily.*

The elephants had found a tiger in the caves which are high on a ridge, so we were faced with a very steep climb. It was the perfect opportunity, in fact, to try out Tuskcam. It was a good test, though we were nervous of the steep climb. There were lots of rocks and trees that he could hit, but he carefully avoided all these dangers. The terrain was impossible for Bouldercam, so Tuskcam was already proving a useful device.

Keyshu guided Mohan so well – the elephant moved slowly and methodically, as if he knew how important the camera is.

The caves are a large collection of rocks, many volcanic, which are near a ridge, high and breezy. It's a cool refuge where tigers can escape from the heat of the day.

This tiger was the male Charger, which we had warned may charge the elephant. As we arrived there was a soft but deep-throated roar, from the entrance of a cave. There he was, looking very cross at being discovered. Without warning, he rushed straight at the elephant closest to him. What an amazing moment! Sadly, I hardly managed a shot in the confusion. It's the most electric experience we've had so far. Such an impressive animal. John was beside himself with excitement on the closest elephant and had done better than me, so we had something on film from this frightening encounter.

Mohan 'wears' Tuskcam like a new piece of jewellery. He accepted the unique contraption straight away as its extra weight makes little difference to him.

Sunday May 22, 2005

OVERLEAF: *Santosh washes Mohan; bath-time is his favourite part of the day.*

Yesterday's highlight was catching footage of a dramatic encounter between a tiger and a Sloth Bear. This morning we made a great effort to be away at 4.30am, as we had great expectations and were full of excitement. We were taken to a new area where again there were rumoured to be cubs.

The rocky, secluded spot looked ideal for tigers. We had two elephants, and Tuskcam was neatly secured to Mohan's tusk. Geoff is enjoying using Tuskcam, and is constantly looking for ways that further improvements can be made.

As we approached the area we could hear occasional calls of what sounded like baby tigers. Then, there they were, two tiny cubs, crawling on the rocks and sliding down them. Enchanting, and all the more remarkable as the mother tiger kept retrieving them and carrying them in her mouth back to the top of the rocks, where their cave must be.

She was obviously comfortable with elephants, so it was as if we were not there as far as she was concerned. We were able to film from Tuskcam and from the top of the *howdah* without any sense that we were intruding.

The mother tiger would watch, seemingly with dismay, as the cubs wandered and slid down the rocks, and she then went down and collected them, carrying them back to the safety of the high rocks.

It was such an emotional morning, as filming the little tigers has given us the start of our story. With a storyline to follow, we'll be able to make the film work. The story has always been key, and has to evolve with a wild subject. Here we have a great opportunity that simply wasn't there when we started, and now could lead us through the seasons.

The cubs make quite a noise when they stray away from the den, but luckily their mother is close at hand to carry them back to safety.

t's been a good week. Two days ago we filmed the mother tiger and her cubs again, playing in the rocks. This time we saw all four cubs. We are having great success with Tuskcam, which Mohan is handling like a professional.

We also found that our remotes, set up by a waterhole near one of the guardhouses, had caught some lovely sunlit footage of a tiger coming down to lie in the water.

Today is nearly our last day for a while, as the park closes for the monsoon. It's overcast, with oppressive heat. It is suggested that we leave the tigress and cubs alone for now, so we'll not see them again this trip. Unfortunately, we didn't find any other tigers today.

We found wild dogs, though. These interesting animals are becoming quite strong characters for the film. We filmed a group of them moving around in the forest and loafing in the distant valley. They had killed a deer in the forest, but there were only crows on the carcass by the time we got there. We followed the dogs to a hand-pump waterhole, where we have put in remote cameras. Two of the cameras got chewed by the dogs and gave a great bit of film. The dogs are habituated to the guard who does the water pumping – he comes daily and pumps for an hour, sitting on a stone while he works – so they were not too scared when I put the remote cameras in.

The mahout's families have come to the elephant camp for one of their regular visits. They stared at us in a friendly but curious manner, adding a bit of welcome colour to the park.

We placed two remote cameras out at another waterhole, in the hope of seeing a tiger but none came. Later, we had a thunderstorm. There was lightning and thunder, but little rain seemed to fall, and it evaporated as it hit the ground, it's so hot. We laughed at how all the monkeys ran away at the sound of the thunder.

The cubs do wander but their mother is very careful to return them to their den in a rocky cave.

Fishing poachers have set fire to part of the park. They had threatened to do this if they were arrested over the long-standing illegal fishing that has been taking place at the dam. The dam (hydro-electric) is huge and a long way from the core area. There are 300 people involved, and they have 90 boats. The park guards are getting a handle on the situation, with 40 boats captured so far. The guards make holes in the boats and sink them – there's a lot of activity under cover of darkness. It highlights the problem of having a large human population near the park; many of these people were displaced when the park was formed so they feel they have a right to fish here. Fortunately the dispute is over fish, not tigers.

It's desolate, barren and bleak with the heat and high humidity, with a real feeling of waiting for the rain, as if everything is holding its breath. Tomorrow we leave.

The tigress takes a rest in the water. Tigers love to cool off in this way during the heat of the day, sometimes spending several hours in the same spot.

ADOLESCENCE

For a young tiger, its first few months are one of the most treacherous periods of its entire life. The ruthless early winnowing of the weak and the infirm is an essential but heart-rending process, particularly when you see it at first hand. On one occasion researchers following a tigress and her cubs noticed that a three-month-old female, smaller than her siblings, was blind. The mother, far from abandoning her infirm daughter, made prodigious efforts to keep her alive but inevitably, she died.

By the age of six to eight months a healthy young tiger is large enough to defend itself against virtually any predator except another tiger; it is also mobile enough to escape the threats of fire and flood, and able to accompany its mother on wide-ranging forays in search of food. The adolescent phase, as with most animals, is the vital stage of learning; learning to survive, to hunt, to interact with other members of its kind and, indeed, with other species. This is also the period during which individual personalities begin to emerge. The critical learning path for adolescent tigers revolves around hunting, through which the skills of the mother are transferred to the youngsters. She will often 'capture' a young deer or pig, trap it under her paw and then release the prey for the cubs to attack and bring down or kill. As with all intelligent mammals there is a degree of instinct involved, but a huge amount of learning is required too and the initial attempts can prove exasperating for the mother, who will sometimes nip her youngsters quite painfully by way of encouragement.

A playful expression from a young tiger. They remain apprehensive during their formative years, only attaining the confident swagger of an adult when they are fully grown.

As skill, size and experience increase, young tigers are exposed to more dangerous prey, including porcupine (*Hystrix* sp.), which probably causes a disproportionate number of tiger fatalities, and Sloth Bear (*Melursus ursinus*). In one hilarious instance witnessed in central India, a pair of cocky male cubs stalked a sloth bear in exhilarated and noisy (on the bear's part, at least) pursuit; a few seconds later the noises changed and out of the bushes shot two tigers in terrified flight from an irate ursine which had obviously realized the lowly juvenile status of its pursuers! In another case, in Chitwan in Nepal, a young male tiger about 18 months old was watched approaching a tethered buffalo. Every time the tiger tried to stalk to an ambush position, the buffalo would wheel round and present an impregnably horned head. This had clearly not been covered in the syllabus, and after several fruitless attempts the tiger dropped down on his haunches, clearly baffled by this unreasonable behaviour.

By the age of 18 months, male tigers particularly are beginning to range widely and mostly alone, although they will often form hunting partnerships or coalitions with their siblings. By the age of 24 months even the female cubs will have left their natal areas, chased off by the mother who will have young cubs by then. The adolescent tigers now enter an extremely vulnerable stage of their lives. They wander widely in search of territory, the males often covering several hundred kilometres in the space of a few weeks, ranging back and forth and eking out an existence in marginal habitats. Here they often come into conflict with humans and as a result suffer casualties from traps, guns and poison. These frontier areas are the sinks into which surplus transients from the productive protected forests disappear, with only a fraction surviving to full adulthood.

Although these itinerant sub-adults will often re-visit their natal areas, and in some cases even consort with one or the other parent, their survival depends upon their ability to establish a territory of their own. Female cubs often manage to find a niche adjacent to their mothers, especially in highly productive areas, but for the males, with fewer niches available, the casualty rate is higher. Conflict with other males and death from injuries or injury-induced starvation is a constant and significant threat.

The mortality rate of tigers from the time of birth until attainment of territorial adulthood is very high – between 60 and 70 per cent, with variations depending on the health and security of the habitat they occupy. One intriguing phenomenon increasingly observed in the well-protected areas of some national parks is the number of large litters – four young – that survive until the age of dispersal, which can be anything between 18 and 24 months. Wherever this has been observed, the mother in question has occupied a prime range, within the stable territorial boundary of a dominant male, and has been an experienced parent – as in the case of the Pench tigress who stars in this book. Given the fact that in the past such a high rate of survival was considered unusual, one wonders whether there has always been an equivalent survival rate of young families and that it is only now being noticed because of the level of direct and long-term systematic observation of wild tigers. Or is the unusual degree of protection and productivity of habitat in some areas the main contributor? Whatever the answer, a tiger that makes it to a territory – especially in the highly competitive environment induced by shrinking habitat on the one hand and healthy tiger reproduction on the other – will indeed have proven its fitness to survive.

Play is a very important part of a tiger's development; it is through play that they learn their killing skills.

Sunday December 11, 2005

We're back after an absence of six months. The park is transformed — such a contrast to the barren scenery of when we left, there's now immense growth everywhere. On our first day out we found and briefly filmed a couple of leopard cubs. Two days ago we visited the area where we had filmed the mother tiger and her cubs in the spring — although no tigers were to be seen on this occasion, just spending time in the real home of a tiger was enchanting. It is a perfectly peaceful setting. The cave itself about the size of the front of a small car, and then a falling slope away to flatter area of woodland at the bottom, where the flattened grass told us exactly where the tigress had been with the cubs. We left the remote cameras in place there overnight.

Today we had decided to get straight on the elephants first thing, and go and look for the female and her cubs. As you'd expect from an elephant, Mohan has forgotten nothing from the spring about carrying us and the cameras. He stood by the car to let us get onto the howdah, so tall he dwarfed the little Gypsy car. His mahout Keyshu wore a blanket wrapped over his head, it's still very cold until the sun is properly up. The creaking of the howdah platform sounded unsafe, but generations of experience ensure the ropes never slip. We sat, mesmerized by the slow plod of the elephant for the 2 km journey through the trees.

It was wonderful. We found the cubs after a little exploration of the den area and it was as if they had never gone away. They were a little timid and nervous, but they were reassured by the confidence of the mother. They were quite difficult to see, as the teak leaves and undergrowth gives them lots of cover. It's a perfect place for them to grow up.

RIGHT: *Geoff Bell rides side-saddle while a mahout has a towel wrapped around his head to keep out the early morning chill.*

BELOW: *Two cubs playing by one of the remote cameras produced an exciting film sequence as they rushed past the lens.*

I was very excited with the possibilities the remotes might deliver, as the mother tiger was sitting in front of one of them when we arrived at the site! She called the cubs out from their hiding place — sheer magic, which we filmed from Tuskcam — enchanting family life, all we could hope for the story.

Thursday December 15, 2005

Yesterday we were lucky enough to find the tigress out hunting first thing. We filmed her catching a fawn, although it all took place in thick cover so the footage wasn't very clear, but it was a very encouraging sign of what was to come.

Today we found her walking down the road just as we were ready to go off on the elephants. We followed her for over a kilometre. It all seemed so easy and matter-of-fact. Whenever I am with a tiger I feel that anything might be possible. She scent-sprayed some trees as she went, one of them the same tree that we'd noticed had been sprayed when I put a camera on the road a year ago!

She went into the forest, scenting trees all the time, pausing and listening as she went. She was clearly hunting. I was a bit surprised as she had caught a fawn yesterday and by the evening she had hardly eaten any of it, saving it for the cubs. Today she caught a chital female. The kill seemed effortless, a dramatic event to witness. The heavy forest cover and the amazing bound of a tiger is a fatal combination for any deer. Even if the deer are aware that there is a tiger in the area they are unlikely to see it until it's too late. The tiger hardly moved at all before she made her decisive move. It was wonderful to be able to film the kill, even though some of it was hidden behind leaves. We took some nice film from Tuskcam of her dragging the carcass to the lantana, where she was to spend the rest of the day.

The cubs feel secure in the presence of their mother; they would only come out of hiding when she was around.

The cubs rely on the security provided by their mother at all times, and they became accustomed to the elephants thanks to her confidence.

It was a good situation for remote cameras, so I placed two there, lowered into place by rope. Each device has a spirit level on its roof, so I can see this from the top of the elephant. There's a chance we might get some useful film.

Shivpujan, the mahout of the female elephant Sarwasti, was keen to teach me the Hindi names of the trees on the way back. This was an impossible task for him, but I tried. It's incredible how many trees have been struck by lightning. The tiger den site is high, and although it looks like paradise now, during a stormy monsoon it must be terrifying.

Friday December 16, 2005

Today was our last day out with the mahouts before we go home for the next couple of months. It turned out that the remotes I placed at the den site had produced some great material of the cubs playing on a log, with their mother coming along as well. Keyshu had told me where to put the cameras, as he could see signs of play on the ground.

It was a good last day, as the tigress came back to the area of the den in the early morning, having walked down the road for half an hour, to the delight of one group of tourists in a car. We were able to follow her and be there when she called the cubs from the rocks.

All of the action was very hidden by the leaves, but we could see how active the cubs are now, running fast to their mother. They followed her as she found a place that was secluded and shaded, so she could suckle them in peace. We took more great film of them playing a little before they settled on her.

They were with the mother for 20 minutes as she let them play on her and then suckle, which finally gave us our first clear view of this. It was lovely to see that the cubs were slowly becoming used to the elephants, encouraged by the total acceptance by their mother. This footage of tiger family life is so valuable for our story. Our confidence is building as we dare to hope we can follow them as they grow up.

OVERLEAF: *The elephants set off with their helpers, who take them to the forest for the night. Chains are attached to their legs which they drag along behind them to stop them travelling too far during darkness. These leave a trail which means the animals can be found easily next morning. The elephants accept them and will often pick them up in their mouths to ease their journey.*

LEFT: *Suckling cubs, a rare insight into the private lives of tigers which we only managed to film only once.*

BELOW: *This enchanting image comes from a series of remote camera shots from the film where one of the cubs suddenly becoming very brave and decided to investigate the camera.*

ADULTHOOD — *the 'King of the Jungle'?*

Adult tigers are huge. The largest individuals are found in south and north-east Asia, where 275 kg (600 lb) is unexceptional for a mature male and larger specimens are not unusual. A tiger shot in northern India reportedly weighed 327 kg (720 lb) and there are reports of a 363-kg (800-lb) tiger once shot in Siberia. Females tend to be smaller than males by a third or so, but interestingly the difference between the largest male tigers and the smallest male tigers from Sumatra (*P. tigris sumatrae*), for example, is greater than the difference between females from the different populations. The whole business of evolutionary size difference is a complex one and animals from the colder north tend to be larger — Bergmann's rule states that lower surface area to body volume ratio leads to thermal efficiency. Add to this the difference in size of prey that males and females hunt and you have another evolutionary factor acting upon body size. For all that, the tiger is indeed the "King of the Jungle" across virtually its entire range, being rivalled in the north only by the Brown Bear (*Ursus arctos*). Elsewhere it fears little, except humans. Leopards (*Panthera pardus*), which share virtually all of the tiger's range, are only a third as large and when encountered are attacked, and sometimes killed and eaten.

For the past few centuries tigers have existed largely in human-dominated landscapes and increasing human intrusion has forced them into the night, establishing and reinforcing their reputation as determinedly nocturnal hunters. There is no doubt that they are well adapted to this role. However, in secure protected areas the tiger is often diurnal. Many of its prey species are either partly diurnal or crepuscular, and it is to the tiger's advantage to hunt while these are active. These national park tigers are unafraid to walk down roads followed by a pack of jostling tourist jeeps, but things change as soon as you step across the park boundary into the buffer zone and non-protected area forests. Here a primary concern is to stay clear of man. In the buffer zone of Bandhavgarh Tiger Reserve — much disturbed by humans and cattle — lived a huge male tiger. He killed domestic livestock regularly but would

Adult tigers spend a lot of their time patrolling and males might walk huge distances in order to secure territories and mates.

kill, eat and then disappear, never returning to a carcass no matter how little he had eaten of it. This is intriguing learned behaviour, which protects the animal from the threat of poisoning or hunter in ambush, and it is interesting to speculate whether or not he passed this on to any of his offspring. Contrast this extremely wary and circumspect behaviour with that of young tigers born within the safe confines of a national park, a considerable number of which will disperse into these marginal forests. One wonders how much this adds to their vulnerability.

For all the romance that surrounds them, the life of an adult tiger is constrained by the iron compulsions of finding food, protecting territory, breeding and protecting or caring for young. Hard graft, in other words. An adult male tiger spends several hours each day doggedly walking the boundaries of his territory, constantly checking the chemical signals that speak of the presence, status and condition of other individuals. The presence of strange tigers must be investigated, and incursions by other males defended, either through increased advertising or physical confrontation. Although tigers are not shy of a fight, injury for a solitary predator can mean death from starvation if the animal is in any way incapacitated. Being large and dominant, tigers prefer to use the most open and convenient roads, tracks, streambeds or

game trails available in the area. Reports of tiger movements from Siberia in the winter speak of them establishing their own preferred routes through deep snow, which become compacted and firm through regular use. Empirical evidence suggests that successive generations of tigers use the same tracks and trails as their predecessors. This pattern of patrolling lends itself to the tiger's preferred hunting technique – finding prey by sight or sound and then stalking close into an ambush position.

Much the same pattern holds good for the female, with the added burden of having to rear cubs. Moving them about – often painstakingly one at a time – when they are very young, hunting enough for all, teaching sub-adults to hunt – all of this falls within the female's workload. Few tigers make it to old age. While there are instances of wild tigers reaching the age of 16 years in the case of females and 12 to 13 years with males, there is insufficient data available to arrive at an average figure. The latter would, in any case, be determined by factors such as quality of habitat and poaching pressure. However, there is empirical evidence that in productive areas old tigers of either sex are tolerated by territorial incumbents, as long as they do not compete for breeding rights.

Few animals have the same beauty and majesty of an adult tiger sitting as though on its throne!

Sunday February 19, 2006

The mother tiger was sighted on the road today for the first time in several days. It is a relief — I was beginning to feel very anxious. It's eight days since we returned to Pench after our winter break and, although the mahouts assured us the cubs are all right, it was a big worry for our story.

Two days ago we found a mother leopard with a small cub while searching for the tiger. Very exciting as we just stumbled upon her. This I knew straight away was something very special, as it's very unusual to see small baby leopards. I immediately lowered a couple of remote cameras down in a likely spot, more or less as we were passing. One of these cameras produced an amazing shot of the mother leopard carrying a cub — I had no idea we would capture anything this good until this evening when we serviced the camera. I left the other camera out on a flat rock in the hope of getting more material. Leopards are so difficult to see, much more so in India than Africa. I was tingling with excitement at having filmed something so extraordinary — I had

to look twice at it, to really believe it was there! It just shows how good the remote cameras are now, as they really are opening up the jungle to us. We are getting a real sense of the world that the tiger lives in.

Today, while following up the sighting of the mother tiger, the elephants found the male tiger Charger, eating an adult Sloth Bear. Another extraordinary sequence for the story. I didn't think a tiger would kill a bear but it seems they will kill just about everything they can. I cannot remember any other film showing this behaviour. It's certainly unusual and it made some interesting film, especially as we managed to use both Bouldercam and Tuskcam. There wasn't a lot of action, as he had eaten about two-thirds of the bear by the time we got there.

I wondered how it had happened, as the bears are so careful and alert. The poor bear was probably wandering along a ridge, perhaps a regular route, minding his own business, when the tiger pounced. Charger was too fat to move and he hardly reacted as we motored Bouldercam quite close to him while he ate. He wasn't going to waste any of the bear.

There was no time left to look for the tigress and cubs, and in any case the presence of the male is likely to have made them hide — if the cubs are not Charger's own, he could well kill them if he finds them, to bring the tigress into season.

Today at last we found the tiger cubs, with their mother exactly where they were thought to be. Such a great relief as it's been more than a week since we arrived on this trip, and of course all worst-case scenarios go through one's mind.

We went in with Bouldercam, just Geoff on Mohan and me on Domini today, as Sarwasti is resting, a new sensible park rule that gives them a day off every six days. We got some great film – the cubs are visibly bigger, the story is beginning to unfold. They were a delight, all in a clearing in the lantana with a dead chital, playing with parts of the carcass. It was great to watch the tigers as a family again, though two of the cubs had been too nervous to come out in the open for a while. As soon as it warmed up they became less active and ended up sleeping in the bushes.

The shots we are getting from Tuskcam are suffering from a bored elephant. Mohan likes to eat virtually all the time. They all do. As they walk, they graze and chew. Their trunks are forever grabbing leaves and branches as we walk through the forest. When he eats he chews, and when he chews he grinds his teeth, and this causes a lot of vibration in his tusks, which shows as a shudder on the film. Even though Keyshu can make him stand motionless at a command, he obviously cannot do it all the time and the whole point is for the camera to film tigers on the move. To solve the problem Geoff has come up with another revolutionary idea – a ball of foam rubber wrapped with gaffer tape for him to carry with his trunk. It's about the size of a football. He will always carry or pick up anything he is asked, so this has been a great invention – whenever there is occasion to film the tigers, either on the move or stationary, Keyshu gives Mohan the ball and this stops him chewing. However, if he has to wait for any length of time he sneakily puts it down, knowing that Keyshu cannot see the bottom of his trunk from the top of the *howdah*, grabs something to eat, and then picks it up again as if he had never let it go, so he has to be watched! The ball is now an important piece of equipment, which comes everywhere with Mohan.

The male cub now shares the kill with his brother.

One of the male cubs, now quite large, is given a freshly killed Chital. He instinctively holds it around the neck in the throttling position.

We found the mother tiger in the same place as yesterday, with a new chital fawn that she was eating herself. Soon after we arrived that she moved off and took the cubs with her on what was to be quite a long walk through the forest.

It gave us some lovely film of her with them as she moved. We took Bouldercam, which works so well now as Mohan is completely confident carrying it. I am also pleased with the way I can confidently film handheld now, sitting cross-legged on the *howdah*. We're getting great shots from both techniques of the tiger family on the move.

The tigress left the cubs at a bush, where they stayed for the rest of the day. They are not quite weaned and are keen to suckle at every opportunity.

She was just about to settle down with them when she heard a rustle in the shrubs. I love it when a cat suddenly becomes seriously interested in something. They become so alert and focused. The tigress stalked briefly, and then pounced very quickly onto something hissing in the grass. It was a pangolin, which somehow survived the attack, as it remained curled up for a long time – long enough for the tigress to lose interest and walk away. We stayed and I was able to film the pangolin uncurling itself and walking away, moving like a mechanical toy. It was remarkable to see, as these animals are normally nocturnal – another great piece for the film. Having had only a brief view of one on the first trip I was thrilled to see another!

Cameraman Geoff (this page) and diary author Michael (opposite) sitting confidently on the howdah.

Saturday February 25, 2006

This afternoon we had special permission to take an elephant out to a remote spot in the river area, a difficult area to reach, not accessible by car. We were expecting to go on the male elephant Jung Bhadhur, but on reaching the elephant camp we found that he couldn't be caught, he was way out in the middle of the lake, dousing himself with water. Nothing was going to entice him back again! So I went with Keyshu on Mohan instead.

We left the camp in just enough time, and it's deceptive how quickly Mohan can cover the ground. There were just the two of us and the elephant down by the wide, deep and rocky river, and all was quiet, punctuated by the calls of the birds and monkeys.

This is perfect habitat for a tiger. We found the prints of a male; the tracks showed he had swum across a pool of water. Who knows when? We carried on in the still air and heat of late afternoon. We found a print very close to a remote camera, probably made at night though; I shall have to wait until later to find out. The second camera also had prints very close to it, perhaps from the same male. It is hard to say, as the sand is very loose there. Mohan struggled a bit up the banks and over the rocks, but Keyshu was very good with him. I looked later at the remotes and found film of the legs of a tiger! No way of knowing which one it is, as it is too close! The remote cameras have a delay on them. When they are activated it always takes a few seconds to start the film, by which time the subject may have moved – in this case the tiger moved right up to examine the camera.

A wonderful shot of a cub emerging from the water and walking right over the remote camera. This angle really does show us what it's like to be deep inside the tiger's world.

HUNTING AND PREY

Tigers will, famously, eat anything. Their adaptability is underscored by writers quoting evidence of tigers feeding on crabs and frogs in the Sunderban mangroves at one end of the scale and on giant 1,000-kg (2,200-lb) Gaur or Indian Bison (*Bos gaurus*) at the other. This simply emphasizes the tiger's opportunistic instincts, because clearly a predator of its size must ensure adequate returns from the energy it expends during the hunt. It has been estimated that an adult male tiger needs about 3,600 kg of meat per year and a lone adult female about 3,000 kg (6,600 lb) – a requirement which sees a significant jump to about 4,500–5,000 kg (9,900–11,000 lb) when she has cubs, depending of course on litter size and age.

While we know that tigers have a very wide-ranging diet, the evidence gathered by the eminent tiger ecologist Dr Ullas Karanth in south India seems to indicate that where easily available, they prefer larger prey – with animals like Sambar deer (*Cervus unicolor*) and Gaur contributing a disproportionate percentage of the annual intake as compared to the much more numerous Spotted Deer or Chital (*Cervus axis*), which averages only about 60 kg in live weight. It would seem logical that a tiger, which has a hunting success ratio of about 1 in 10, or even as low as 1 in 20, would prefer to concentrate its energies on larger prey – more bang for the buck, as it were! However, it is usually the males that kill animals the size of Gaur, while females will generally stick to smaller prey. This does not mean to say that females will baulk at taking on prey larger than themselves; in Chitwan in Nepal a tigress was once watched in a two-day hunt of a female Gaur, having first hamstrung the animal and then returning the subsequent night to finish it off, albeit after a tremendous tussle – as evidenced by the state of the surrounding vegetation.

Inhabiting woodland and high grass habitats, it is clearly advantageous for a tiger to hunt alone, finding prey as it quarters the jungle and then stalking it. Tigers hunt primarily by sight, aided by sound and scent, and ideally the approach to potential prey will be made to a distance of about 20 to 30 metres, the final distance being judged exactly by moving the head backwards and forwards a few times and then delivering a short, powerful charge. Generally speaking, if the prey is not caught within a

A female Sambar is caught in a small valley. All deer are vulnerable when they look for water in such places.

distance of about 100 metres at the most, the tiger gives up. There are exceptions, of course. A famous tiger in Ranthambhore would chase Sambar into the lakes over a considerable distance, making a remarkable success of this technique.

If a charge is successful, the actual kill then becomes a critical test of strength and skill. As the tiger knocks the prey down, it uses its massive canines to deliver either a crushing bite to the neck or, more commonly, a choking bite to the throat, which it needs to hold for several minutes. Throughout the process it must avoid flailing hooves, antlers and, in the case of large boars, razor-sharp tusks. Every kill is therefore a challenge and it is no wonder that male tigers will happily annex the kills of female tigers or of other smaller carnivores like Leopards (*Panthera pardus*), scavenging or even eating putrefying meat.

Having made the kill, a primary concern for the tiger is to drag it into cover. This is not only to protect it from other tigers and scavengers, but also – going by the evidence gathered by Dr Raghu Chundawat in Panna in India – to shelter the tiger itself from the sun and the risks of overheating. Tigers will most often begin

Male tiger eating a male Sambar, which is India's largest deer species.

feeding from the rump, while leopards first open the stomach – a useful but not infallible way of identifying the kills of these predators. The hair is removed and the skin and flesh cut with the carnassial teeth. Once the stomach of the prey is opened the tiger pulls out the rumen sac and sometimes drags it several metres from the carcass. Tigers can remain at a kill for three to four days and hungry male tigers have been known to eat up to 35 kg (77 lb) of meat at a sitting. Tigers are pure carnivores, eating only meat, which is easily digestible and requires only a short gut.

A romantic notion of tigers held that they killed only to satisfy hunger. This is not true. The Bandhavgarh tigress Sita was once watched sitting on a fresh Chital stag kill when another stag wandered within range and was promptly despatched; and in Chitwan a male once killed several head from a herd of livestock feeding at the edge of the forest. In neither of these cases did the tiger concerned feed on more than one kill. Making a kill is always a challenge, so it seems natural that no opportunity is passed up.

Wednesday March 1, 2006

Today was a good day as we found the tigress early on, walking down the road along a familiar route. It was interesting to witness her leaving the pugmarks, which will excite everyone when they later drive down this road.

I travelled on Sarwasti, while Geoff rode on Mohan. We both set off, filming with our different methods as she scent-marked trees. She would sniff and then lift her tail and spray urine forcefully at the tree trunk. It's a great image, really establishing that this is her patch, though the smell is actually quite noxious to our noses.

The cubs are nine months' old and the mother is still spending most of her time with them.

She looked for deer as she scent-marked. There were very few deer here, but with a sudden acceleration she caught an adult Chital. It all happened very quickly some way in front of us, so by the time we arrived, we could only film her dragging the carcass. We had been struggling to keep up with her, as she bounded ahead of the elephant.

It felt good to have got some more material from her, and when Bouldercam arrived with Mohan, we got some quite intimate film. She was eating a bit and then sleeping again. No sign of the cubs in the morning, so we left Bouldercam there and gambled on her staying there until we came back in the afternoon. This paid off – when we returned she and her cubs were eating the deer. Bouldercam's footage is amazing – I feel I am lying next to them when I see the film. One of the four cubs is still very shy so it's really difficult to get them all together.

There was thunder this evening and a little rain.

The tigress shares her kill with the cubs, who are now very keen meat-eaters.

Unusually, we awoke to thunder and lightning at 5am when my tea arrived. There was not much rain. We found the tigress and the four cubs early. They were very sleepy and fat — they must have caught something after the rain last night. The shy cub remained as elusive as ever.

The tigress still had an untouched Chital fawn and a partly eaten carcass of an adult, which the cubs did show a little interest in. It seems likely the tigress caught the mother first and then easily caught the confused fawn.

One of the cubs was hauling the fawn up a rock, and this made a great piece of film on both of our devices. It shows how strong the cubs are becoming. The growth of the cubs is such an important part of the story, so this is useful stuff. The mother tiger remained nearby in the rocks, so it was a good static situation for Bouldercam, which does produce a real sense of actually being there with them, as if the viewer were one of the group.

Bouldercam films the mother and cubs as they eat another deer. The cubs were very relaxed with the camera and the mother didn't seem to notice it at all.

Digpal is now with us as our new guide. He is our liaison with the directors of the park. He is very diplomatic and speaks Hindi, which we cannot. Things are changing a bit and it looks like we cannot use the elephants in the afternoons any more — a real shame as the last two hours of the day can be magic. It's understandable though as the days are hotter now and the elephants need more rest.

Wednesday March 8, 2006

Yesterday there was heavy rain and the elephants decamped to the forest for the whole day. The tracks were too waterlogged for vehicles, so we were limited to filming a local jackal family, who provided some entertaining footage. There was no more rain overnight, though, and it looks like it will dry out now.

We found the tigress and cubs this morning – the mahouts had heard her from their camp, so went straight there. We were able to spend a lovely morning with the family.

I captured an extraordinary piece of film of the tigress hunting in the lantana. The bushes are so thick that there is no visibility at all for any large mammal that might wander into it, but the deer do seem to like to be here. I guess they feel very well hidden and therefore safe. They must have to build up an auditory picture of what is where in these jungle bushes. Anything that moves can be heard – there is plenty of dead leaf material on the ground to turn any footstep into a crunch.

Suddenly the tigress was up and alert, as there were three Sambar deer walking through. She can stalk very silently, and she began to creep towards the Sambar. They were quite unaware that a tiger was there. She must have been completely invisible.

We could hardly see the deer, but we knew she could, as she moved very slowly forward. I was desperate not to lose sight of her, as I strained down the viewfinder. Then suddenly the Sambar were aware of the tiger, and they panicked and ran. The tigress went for one of them as it ran straight towards the elephants, as if we weren't there. She sprang upon the deer right in front of my elephant, so fast I hardly saw it. The elephant was so surprised that I think she thought that the tiger was coming for her. As she backed off she nearly sent me flying. Such drama!

The replay of the film revealed something amazing. Somehow I'd captured the impact of the tiger and the deer right by the elephant. Extraordinary!

The cubs appeared very quickly, as they must have heard it all. We were able to get Bouldercam quite close and obtain more material of the tigress with the cubs – Bouldercam can get right under the lantana bushes, and produce an image that is just impossible to see from way up on top of the *howdah*. Eventually they all went to sleep and we had to leave. We were able to drive Bouldercam away out of the bushes and back to Mohan without the tigers moving at all, not one of them lifted its head. A good morning.

By evening there was a great electrical prelude to a thunderstorm.

The female tiger bounding after a deer. It was always a great thrill to watch the tiger's athletic abilities when hunting.

Sometimes the stalking tigress would walk right up to our elephant, giving us an extraordinary view. Here she is looking intently at deer right below us, then literally flying at the potential prey to try and force them into an error.

OPPOSITE: The cubs have been left on their own while the mother goes hunting. They became quite shy and nervous without her, remaining together but not making any sound and not wanting to engage in any play among themselves.

OVERLEAF: Diary author Michael prepares for the rain before setting off on Domini with Sadahan. The rain was often intense so we had to keep the equipment dry.

Thursday March 9, 2006

Another wet morning, but it was dry enough to get on the elephants early. They had found the tiger. The mother and her cubs were in the same place as yesterday. Not so surprising, as they had their kill there.

We got Bouldercam into position and filmed the tigress in a thunderstorm. It felt like quite an achievement even though the tiger, of course, did very little except sit it out. We captured nice images of water running in beads off her back.

Each elephant had an enormous tarpaulin that covered both the mahout and ourselves, so we could film throughout the storm. I kept thinking of all those trees around with lightning damage – the lightening and the thunder seemed dangerously close, the flash and bang happening virtually instantaneous. It was just a little bit scary. Each flash of lightening made the elephant's body jerk as if it was experiencing an electric shock. They are so sensitive – it is impossible to know how they really feel in such a storm, but I felt very vulnerable. I could begin to understand why the elephant assistants are reluctant to go out into the dark forest in such storms to look for the elephants when they wander off.

Friday May 19, 2006

We are back after a couple of months' break, and it's really hot now. Today got off to a good start as we got straight on the elephants. The mahouts knew where the tigers were likely to be — fantastic news as it is always an anxious time when we return. We found the mother and four cubs all together very quickly. Suddenly it all seems easy again.

She had killed a female Sambar, and had finished feeding by the time we arrived. The cubs looked bigger again — it's good to show this progression in the story. They were lying in the shade of the riverbank on cool, damp sand. They ate a little more and then went to lie in the water. Absolute magic, this was the first time that I had been able to film all five of them together as even the shy one was there, now that much braver. They look bulky for their size, slightly ungainly.

As it's so hot now we didn't have that long with them. It takes the best part of two hours for the elephants to get to the dam, where Digpal was waiting with the car for us. John is very excited that BBC controllers have commissioned three one-hour programmes on the strength of an eight-minute promotional DVD of our successes so far. They recognize that it will be a great story of tiger cubs growing up in the jungle. By the sound of it they seemed genuinely amazed by what they saw. It's so good to see that the film is starting to work — but the expectations of what we can achieve are now so much higher. We really will have to get something wonderful with our family now!

Two cubs lounge on a rocky outcrop; they couldn't look more comfortable despite the surface upon which they are resting. The tigers seem to enjoy the coolness of rocks that are in the shade.

Habitats and Corridors

For all the dangers of poaching, the most pernicious threat to the tiger remains the disappearance of its habitat. Generally speaking, tiger habitat is forest, the context within which natural selection has honed to perfection a specialist stalk-and-ambush predator of closed country. However, ideal tiger country is not the dense tangled jungle overgrown with vines and creepers of popular imagination; rather, it is mosaic forest consisting of a patchwork of closed and open forest interspersed with productive grassland, where the nutrients are close to the ground and available at different levels to a wide variety of the large ungulates that form the tiger's primary diet.

'Productive' habitat is the kind of country where ungulate numbers, especially of the larger deer, buffalo and wild oxen, thrive and in turn sustain high tiger densities. A classic example of this habitat occurs in Kaziranga National Park in Assam, where the alluvial flood plain of the Brahmaputra River is covered by tall grasslands interspersed with dense riparian forest and semi-evergreen forest in the contiguous hill country. It is a habitat that supports high numbers of Asian Elephant (*Elephas maximus*), Great One-horned Rhinoceros (*Rhinoceros unicornis*), Wild Buffalo (*Bubalus bubalis*), Swamp Deer (*Cervus duvauceli*), Wild Boar (*Sus scrofa*) and the ubiquitous Hog Deer (*Cervus porcinus*). Unsurprisingly, Kaziranga also has probably the highest tiger density in the world – about 20 tigers of all ages per 100 sq km (39 sq miles). The park exemplifies the incredibly rich *terai* habitat, the dense jungles and alluvial grasslands of the sub-Himalayan foothills and narrow adjoining strip of sloping plains that once ran for 1,500 unbroken miles along the length of the mountains and formed one of the richest and most productive wildernesses in Asia.

Seventy years ago the supply of wildlife here seemed inexhaustible. One Nepali hunt bagged a 120 tigers over a month and within a couple of years the tigers were as numerous as ever. A crucial factor was the vast swirling spread of jungle – shifting, thinning, thickening, accommodating small islands of human habitation – and the corridors that connected them, allowing the surplus of any sub-population to quickly repopulate an area that had been emptied of its own tigers. In this regard it is interesting to note that human modification of the habitat in the past, primarily through slash-and-burn agriculture that opened up forests and left fields fallow for long periods, allowing for succession vegetation of grassland bamboos and shrubs to colonize the open areas, acted greatly to the tiger's benefit. Indeed, famous tiger reserves like Kanha and Bandhavgarh have been greatly enriched by the fields of the villages that were relocated out of the parks, leaving behind the wildlife-filled meadows around which tourists now cluster.

Elsewhere in Asia, potentially good habitat for tigers survives but has been stripped of much of its wildlife by uncontrolled hunting. Less than a century ago the alluvial flood plains of great rivers like the Mekong, Irrawaday and Salween supported vast concentrations of wildlife, including large ungulates. Tigers were widespread and common. Yet today, many of south-east Asia's forests are virtually empty of anything but the smallest mammals. No surprise, therefore, that with little in the way of prey to support them, there are perhaps no more than 150 tigers left in the whole of Thailand.

When the forest was dominant in Asia it supported perhaps a 100,000 tigers. Today a sea of humanity laps and eddies around the wilderness, eroding and fragmenting it. What remains is further devastated by poaching and, in India, by the hordes of scrub cattle that are kept for manure, draught energy and milk. These drastically reduce the carrying capacity of forests, competing with wild ungulates for forage, trampling new growth and hardening the soil. All too often they fall prey to tigers, leading to conflict with the owners. The isolation and degradation of small patches of tiger habitat is a serious threat, because in this situation the reduction of a small population through poaching or disease becomes irreversible, with no chance for the surplus of another population to recolonize the emptied area or revitalize the diminishing numbers.

Tragically for the tiger, we also now live in an age of giant infrastructure projects. India, with its galloping economy, is busy disrupting vital habitats with high-speed roads slicing through key tiger habitats like Buxa Tiger Reserve on the Indo-Bhutan border and the extensive tiger reserves of central India. In remote country infrastructure projects bring roads and all the perils therein. A study in the Russian Far East demonstrated that the creation of a road doubled the mortality rates of tigers. In northern Myanmar the wildlife biologist Dr Alan Rabinowitz travelled through the Hukaung Valley in 1999 and was amazed at the wealth of wildlife and tigers that survived. Barely four years later, the revival of the Ledo road had resulted in such an erosion of this wealth that a camera trap count estimated a tiger

Tigers can live in places with surprisingly sparse vegetation, meaning that any forested area has the potential to support the species.

population of just two per 250 sq km (100 sq miles), as opposed to the figure of ten which Rabinowitz e had expected.

Tigers are wide-ranging animals requiring the largest possible area to roam around and disperse into. In recognition of this, new conservation strategies stress landscape protection through which key protected 'core' areas are linked to each other through 'corridors' – multi-use forests in which humans also operate. The Terai Arc Landscape project (see page 121) is testing and fine-tuning this approach to tiger conservation and will hopefully serve as a model for application elsewhere.

Sunday May 21, 2006

Yesterday we filmed the tiger family dozing and got some nice footage of a couple of the cubs playing in the water, so we went back to the same area of the forest again. No tigers, but pugmarks on the road showed that the tigers were near.

We followed the tracks and soon found the tigress lying in the bamboo forest above the dam. The kill from yesterday was finished. It's so hot now, 44°C (111°F) on my thermometer. She doesn't wander far from water and today is no exception.

Rustling in the dry bamboo gave the cubs away. The leaves make a perfect new bed for the young ones wherever they decide to lie down. Filming was tricky with the remote camera devices. Geoff managed to get our new device, Trunkcam, close enough to the sleeping cubs in the dry leaves to get some enchanting footage. I also managed to lower a remote camera from the elephant to film them as they moved from one shady spot to another.

Cool cat: the mother tiger in one of her favourite places to see out the heat of the day. It is a well chosen spot as she not only has the relief of the cool water, but also the shade from the tree above.

Later the tigress moved to the big tree in the dam. This must be a favourite place as she had settled there to shelter from the sun the other day, but then without the cubs. This tree gives some deep shade under its fallen trunk that stretches from the bank right out over the water.

The tigress brought her cubs there today and they crammed themselves together in a fight for space in the shade. They gingerly touched the water with their paws as if they hated it, but then sank into the shallow water at the edge of the lake. They lie there, apparently in bliss.

Again it was tricky to use the remote devices but we managed to get some good bits from Mohan with Trunkcam. It was so rewarding to be with the cubs as they relaxed in the water.

Here the cubs also discover the benefit of the shade of the same tree. They gradually learnt that this was a good place to be and were happy to play in the water away from their mother with increasing regularity.

Monday May 29, 2006

One of the female cubs; always listening, she was alert to all sounds.

Until today, the tigress had not been seen for more than a week. It is possible that she has temporarily left the cubs in order to lead away an intruding male. The Forest Department instructed for three *machaans* (tree platforms) to be built, from which the guards could watch for her return. The park was understandably taking her disappearance seriously.

Today, to our relief, the tigress was found. She was heard calling in the woods close to a spot where the cubs had been seen yesterday. We could still hear her from the road, even before we got onto the elephants. We were allowed to get straight onto the elephants and followed her calling. We found her where we had looked yesterday. There was no sign of the cubs though, and in the still of the early morning with her calls carrying so well I was surprised that they did not appear. She didn't seem too bothered, so neither did anyone else. We settled down with her by a rock without the cubs.

As she lay down in the shade of a large rock, Keyshu spotted a leopard on the rocks above and beyond us, so I managed to get a little bit of film of him. He looked at us and then walked slowly away.

OVERLEAF: *Young tigers are very inquisitive and will often climb up into trees as they explore and play.*

BELOW: *The male leopard had obviously eaten a large meal as he was looking a little weighed down, but there was no evidence of what it might have been. He tolerated us for a short while and then sauntered off the way that leopards do.*

Tuesday May 30, 2006

No mother tiger today but we weren't too concerned. We found the cubs near the cave where they spent their first few weeks of life. They must know the area well and it's not that far from their favourite waterhole. They were sleeping and I was finally able to retrieve the remote cameras we had left by the water five days ago.

The remote cameras had captured some great film of the cubs playing with sticks in the water, three of them very nicely framed, and for some time. Then in the same position there was a leopard coming to drink. It told a worrying story for if the leopard had found the cubs alone he would quite likely have killed them.

As if this weren't enough, on the other camera was film of an adult male tiger! Just incredible. This is easily the most productive deployment of the cameras so far.

The male tiger in the water was a big surprise, as we had no idea he was around. It's one of the exciting aspects of these remote cameras. This may well have been the reason for the mother's long absence — the mahouts had theorized that there could have been a male around that she was trying to lead away from the cubs. All this is showing just how vulnerable the cubs really are — will they survive to complete our story?

ABOVE: *Leopard caught drinking by the remote camera.*

RIGHT: *The cubs would often play in the water with anything they could find, in this case a stick, with their exploits being recorded on one of the remote cameras.*

Saturday June 10, 2006

Rain plays a big part in these elephants' lives. Their helpers are fully equipped as they take the animals off to the forest for the night.

t's our last full day before the park closes once again for the monsoon. I was out and away by 5.45am this morning and away to look for a pair of mating tigers. The mahouts had heard calls all through the night, as so often happens – tigers pair up and mate repeatedly during the female's receptive period, which lasts several days. There were tracks down some of the road and they thought possibly it was Charger and another female. It all looked exciting, but alas came to nothing. It then took us two hours walking at the slow elephant plod back to the cubs, which had been seen from the forest road at one of the dams. Geoff was there with Mohan and two cubs that were sleeping in front of Bouldercam when I arrived. There they remained, wanting only to sleep in the shade.

There was heavy rain in the afternoon, which was all rather frustrating as we had no access to the elephants and all we could really do was to drive round and pick up the remote cameras. It was fitting in a way; as soon as it rains the possibilities offered by the waterholes vanish. Suddenly there is fresh water everywhere, and the puddles on the road are enough to quench an animal's thirst.

Yesterday afternoon we checked the nine remote cameras that we have out now. As they can stay out for days at a time, and no longer fire off at night we have begun to get some great material – it really feels like we spying on a secret world. They are all centred round water, as the heat dictates that animals must drink sometime during the day. Insects, particularly bees, are triggering the devices so some tapes show just blank scenes, but others are truly rewarding. I place them with great care now, so that insect flight paths are avoided.

Some of the images are simply enchanting – something that an artist might paint, they are picture-book images as the animals approach and then relax. The tigers today all round the camera were quite lovely; no one will be able to resist looking at them.

The Chital (or Spotted Deer) are capable of running up to 40mph to evade predators.

93

COUNTING TIGERS

ies, damned lies and statistics, Winston Churchill is once reported to have growled in frustration, and confronted by the baffling variation in tiger estimates, the splendid diversity of numbers embedded in tiger lore and the fervour with which one or other number is seized upon to further a particular argument, one may well be inclined to echo him. So, what is the basis of these numbers? How were they arrived at? How credible are they? And finally – does counting tigers really matter?

The answer to the last question is self-evident. Numbers are the essential, basic data that inform and influence all conservation decisions – planning, policy and long-term strategies. Numbers allow one to assess success or failure. But is it necessary – or even possible – to count every single individual across entire regions? Counting tigers is no easy task. This is a shy animal, at its most comfortable in darkness and deep cover. Counting by sight, and identifying individuals by their body markings, is therefore ruled out. Tigers do, however, leave signs – especially tracks. The time-honoured census method followed in India, therefore, is what is known as the pugmark count. This is based on the fact that within a limited population individual tigers can be identified from the variations in their tracks. Even within a very limited area, and dealing with a small population, the pugmark count relies on the skill, experience and integrity of the person or persons collecting and identifying the pugmarks. Furthermore, and just as critically, it also requires the widespread availability of a suitable substrate to accurately reproduce the size and distinctive individual variations in a consistent way.

Even when these criteria apply, the census exercise must be a sustained year-round effort and not a simple annual or biannual affair. To expect such conditions to be replicated across many thousands of square kilometres of tiger habitat, involving huge numbers of workers, many of whom are forestry employees with little or no experience of tigers, let alone their tracks, is clearly not realistic. The fact is that the results of every census held in India must therefore be treated with scepticism.

Numbers hold a fascination for humans. We see them as compelling 'facts' and, in a field as arcane as the counting of tigers, have traditionally rarely questioned their credibility or relevance to the business of conserving tigers. Even the ongoing census in India will be of little practical consequence unless it accurately maps the expansion or contraction of tiger range as well as quality of habitat and associated tiger densities. This will hopefully obviate anomalies such as the high tiger densities in low-quality tiger habitats which are unquestioningly entered into official reports.

Whilst ecologists in Siberia are attempting to evaluate tiger populations through site-specific sampling techniques during the winter, these are designed to measure actual tiger numbers against potential carrying capacity based on the estimated prey base, rather than to establish absolute numbers on a regional level. The countries of the Indian subcontinent are the only ones that attempt nationwide tiger counts, the practicability and relevance of which is questionable. Tigers were once abundant across the whole of south-east Asia, a region that even today supports far more extensive and potentially productive tiger habitat than India. Yet, rampant hunting pressure has reduced these forests to canopied deserts, emptied of their faunal wealth and thus their wherewithal to support tigers. In such situations the most fundamental information is simply to establish the presence or absence of tigers and map the extent of their range accordingly.

Such information can be gathered by simply recording the occurrence of tiger sign, be it tracks, scat or other evidence. With regard to the actual counting of tigers, India's most eminent tiger ecologist, Dr Ullas Karanth, has stressed that the most important information required for the management of tiger habitats is to assess the carrying capacity of a particular area by undertaking accurate prey base census using well-substantiated sampling methods. The known cropping patterns of tigers will yield a reasonably accurate number of tigers that the prey base will support. It is also important to know the population trends of tigers. Dr Karanth notes that this can be achieved by well-designed sample surveys of tiger signs, which yield indices of signs encountered over set transects. Such indices can be used to measure the population trends.

However, accurate counts in specific areas are important for management purposes and to corroborate simpler sampling methods. For this purpose Dr. Karanth and his colleagues have adapted a technique known as capture-recapture sampling, in which tigers are photographed using trip cameras and identified through the unique patterns of their facial and body stripes.

Using photographs taken over set periods, a capture history of each recorded tiger is compiled and then analysed using computer models. This yields estimates of the captured tigers as a proportion of the total population (not counting cubs), thereby allowing calculation of the actual on-ground total with reasonable accuracy. This allows ecologists to measure *real* tiger numbers against the potential carrying capacity of specific areas. The importance of this information is paramount, as it informs all management and protection decisions.

If the tiger is to be rehabilitated in even the most modest measure across the remnants of its former range, it is vital that

Making a plaster-cast of a pug-mark to try and identify an individual tiger. Counting tigers is essential to tiger conservation.

achievable conservation goals are set. Used judiciously, a mix of census tools should yield the information required to help prioritise the following key goals: which specific areas amidst the expanse of south-east Asian jungles need to be given protection; where and how does habitat need to be managed to optimise productivity; which are the corridors that can and should be secured to connect sub-populations; and where should limited resources be most effectively spent?

t's our first day back after the monsoon – the park only opened a week ago. Everything is cloaked in a profusion of green growth – once again we're struck by the incredible change. The cubs are four months older now. I was asked to go to meet the Director at Seoni, the park headquarters, which is an hour's travel away. I went in the late morning and ended up talking to him for two hours, mostly about the biology of the reserve as he is a biologist. He has managed to catalogue over 800 species of plant, and photograph them too.

He showed me part of his incredible photography collection of all animal life on his laptop. He has published a field guide for Pench which is a great help to the rest of the staff.

He also showed me video and stills of a dead male adult tiger. He had done an autopsy on the to establish the cause of death – everyone had assumed the tiger was killed by a poacher's gun, but in fact he had died fighting another male tiger. This was clear from the lacerations to his body from the claws of the aggressor, and tiger fur in the victim's broken claws. There were canine holes in the neck, in the jaw and through the tongue. The neck vertebrae had been broken, which was the ultimate cause of death after a great struggle. The body was weighed and burnt on a funeral pyre. It's a sad story but an interesting part of tiger life, showing that the greatest threat to tigers is often other tigers.

These giant orb-weaving spiders of the genus Nephila *are everywhere in the forest. They start off with tiny webs at the beginning of the monsoon and soon grow to a size that will span the palm of a human hand. We become entangled in their webs everywhere we go at this time of the year.*

FOLLOWING PAGE: *Shivpujan holding a wand full of spiders' webs. All the mahouts use branches from passing trees to make a wand to gather the webs so that they don't get in peoples' faces all the time. The spiders end up all over the elephant and have to be brushed off.*

Monday October 9, 2006

Our first ride out in the early morning today was like entering paradise; such a change from when we were last here. Dragonflies and butterflies were out in profusion, and looking to capture them were a maze of spinning spiders, their webs spanning the road in some places. Seeing these amazing structures through the trees, it would seem impossible for any flying insect to survive here now. The spiders, each the size of a human palm, are forever catching in my hair. The mahout holds out a branch like a wand and waves the webs away, as we march onward. The elephant's eyes get completely covered with spider silk and lost spiders run back and forth across her head. She's not bothered but I find it all a little disconcerting.

I was very excited at the prospect of seeing the family of tigers again, as the mahouts saw them, all together and well about a week ago. It's another great relief for us, as anything could have happened over the three months of the monsoon.

We saw nothing but a few deer this morning, though. We went up to the caves where we had seen our tigers in the past but the caves were all overgrown and unused – we will have to wait a bit longer to see the tigers.

It is such a hidden forest now with all the growth. Water is still very much in the atmosphere in the morning – it's dripping from the trees and there are still babbling brooks at the bottom of the valleys. It's very difficult to decide where to put the remote cameras.

We spent the afternoon trying a newly improved Tuskcam on the infamous Jung Bhadhur. We are keen for him to be able to carry the device so he needs some training to get him used to it, as this will give us an alternative to relying on Mohan all the time.

Because of his bad-tempered reputation we were a bit nervous. The only other time we had tried anything on his tusks he had, after a theatrical pause, neatly and firmly kicked it off. But today he was very good, quietly accepting the device as Geoff slid it over his tusk.

We left it on him for a while and got him to walk round with it. Then, to mimic the camera, we used a dummy device made from a bucket, which allowed him to feel totally at ease carrying an object on his tusk.

Mohan patiently waiting with Tuskcam, which now fits seamlessly to his tusk.

We had a second successive great morning with the tigers today. Keyshu found them in good time and John was thrilled to see them hunting. They were all there, but the cubs were hanging back, which they seem to have learnt they need to do if their mother is to stand much of a chance. As she walked she continuously looked for prey and stalked some Chital.

There were several dozen deer scattered in the forest and one of them must have seen her, as there was suddenly a chorus of alarm calls very close. We were with the tiger and could see what was happening. Not all the deer could see her and as soon as she started to stalk she became even more difficult to see. It must be instinctive for the deer not to move very far when they know there is a tiger in the area, as they huddle together and stare at the last place any of them saw the tiger. They are very reluctant to go into a *nyala*, as they must know that it presents a high risk of ambush.

OPPOSITE: *Geoff Bell approaching the cubs with Tuskcam on Mohan. The camera is invisible to the tigers as it looks like part of the elephant.*

BELOW: *Female Chital in one of the large harems that form in Pench at this time of year.*

One of the cubs sits and waits for its mother to return. They often get left for long periods and during that time they will sit and listen for her to call, which she will do when she returns whether or not she has been successful in hunting.

The tiger crept very close to the deer, and adopted a frozen form, relying on her cryptic camouflage until the very last moment, hoping that the deer would move towards her by running through the *nyala* and giving her a chance to pounce as they run through the sandy gap.

It wasn't to be, as they went the other way. Watching the deer using their senses for survival is a marvel, I am sure if I were in their situation I would panic into her jaws.

It is interesting watching much of all this through my black-and-white viewfinder. The tiger, which normally is very visible with her bright colour against the greenery, is virtually invisible in monochrome which is, I suspect, closer to how the deer perceive the world.

All this gave us some fine moments for a hunting sequence, and if this was all we got today I would have been happy. But then just before we were due to leave one cub started to go into the water. Fantastic! It's encouraging to see how the film is coming together. It's beginning to exceed our expectations, which is a wonderful thing to feel we can say!

Trunkcam filming the cubs, which are all sleeping after a heavy meal. They are able to eat huge quantities of food in one go.

103

Friday October 20, 2006

There were tiger tracks very close to where we got on the elephants. I find it difficult to distinguish a female's pugmark from a male's, but everyone I am with seems quite certain of what we are seeing in the sand.

These tracks led us to the mother and the cubs; they were looking alert and went hunting. The cubs, which are now approximately thirteen months old, all followed their mother as she stalked the deer.

Two male cubs totally relaxed after a meal. They will stay together until they are fully adult, then have to go their separate ways in order to find their own females.

104

They still have a lot to learn and she provides their only education. The cubs look big, almost adult, but this is deceptive as it's clear by the way they behave that they still have a lot of growing to do. They followed the tigress constantly, running to greet her when she stopped. But it was still difficult to get shots of them all together.

They made a kill, but it was unseen and well ahead of us, as can so easily happen. In fact, it was two kills, both chital and one with a foetus, which one of the cubs was playing with as a cat would with a mouse.

We introduced Trunkcam to the tigers and there was no reaction to the new device. It was able to get really close to the tigers as they took turns at the carcasses. The cubs have become more aggressive towards one another – this will eventually separate them completely, as they become independent tigers.

Saturday October 28, 2006

We found all five tigers in the area where they had been seen last night. Geoff was on Jung Bhadhur with Tuskcam, which works fine, as he appears to hardly notice it, and I rode Sarwasti with Shivpujan. We had a really rewarding day – it was great to find the cubs early.

Almost immediately they started to play with each other. They spent a lot of time stalking each other, which is entertaining, as the 'victim' appears not to see the stalker until the very last moment. This is how it will be in later life, when they come across a tiger they do not know. Their lives will be mostly solitary so they need these skills.

The tigress came and called them away, giving us some lovely shots of the cubs jumping on her, something I have found quite difficult to get on camera, as they will normally run ahead of us whenever she appears. It is often all over by the time we get there.

The mother tiger then went off on a mission to look for deer and all the cubs came too. It was different today as she was letting them run ahead. It looked like she was using them to flush out and panic the deer, as the cubs stood no chance on their own of catching anything, but the deer react with alarm at them just the same. The mother stayed back, watching the deer as they regrouped and dispersed. I think she hoped for one to come straight to her, which must happen on occasions, but not today. The deer must have been confused at seeing more than one tiger at once, probably quite alarmed. You might think that the deer wouldn't stand a chance but in fact the young tigers are very chaotic in their excitement at seeing the deer, and are very obvious to the deer as they lollop carelessly through the forest.

Tigers spend most of their day keeping cool in the shade, often near streams and ponds.

THE CONFLICT BETWEEN MAN AND TIGERS

Tigers and humans have always been in conflict, simply because they compete for identical resources – protein, and the forage and browse needed to support an optimal prey base on the one hand, and livestock on the other. The latter has always supplemented wild prey in the tiger's diet and the predation of livestock remains one of the most intense areas of direct conflict to this day, leading to increasingly lethal retaliation by humans as they gain increasing access to cheap and deadly chemicals and weapons. Until the widespread availability of efficient firearms at the end of the last century, the tiger was dominant in the area of direct physical conflict. Indeed, some authorities believe that until recent times tigers may well have viewed humans as legitimate prey. The fearsome depredations of man-eating tigers upon the early settlers of Singapore – and the terror this inspired – is a matter of record.

Clearly it was not all one-way traffic, and human hunters contrived a variety of methods to trap, spear, net, snare, poison and shoot tigers. Nevertheless, the pressure on tigers from hunting remained minimal until the advent of firearms. Then, as now, the real conflict was played out in the struggle for living space, and in this domain human technology – fire, axe and plough – dictated the terms. Even so, the shifting agriculture practised across so much of tiger country by a sparse human population almost certainly benefited the tiger by modifying the habitat in favour of large ungulates. As human populations increased, the conflict between the wild and the settled intensified. In the vast tracts of the south-east Asian rainforest, teeming with the large prey so favoured by tigers, the spread of modern firearms unleashed an orgy of sport and subsistence hunting that so effectively decimated the tiger's prey base that prey species such as the Kouprey (*Bos sauveli*) – only described to science in 1937 – have today been reduced to a far more precarious situation than even

A traditional tiger hunt in the early twentieth century.

A sad sight: many tigers are still kept in such cages and used in circuses.

the tiger. In China and India particularly, exploding human populations devoured vast tracts of the richest jungles, such as the incredibly productive sub-Himalayan forests of the terai.

During the nineteenth century British colonialists — seeking to recreate in India the artificial countryside and managed game estates of their homeland — had tried to eradicate tigers as vermin. But so resilient is the species that, with the prey base intact, it saw out that threat as well. Not so in Maoist China, however, where all wildlife was seen as competition and tigers and their prey base systematically eliminated as pests inimical to human well-being. The result is that only a handful of tigers possibly survive in southern China. The heightened demand for tiger parts as ingredients in Traditional Chinese Medicine (TCM), and the pressure this has exerted on drastically diminished tiger populations, is too well known to bear detailing. Tragically, just as this demand was sought to be curbed by a finely calibrated blend of international pressure and domestic reform, another menace emerged in the form of expanding demand for ceremonial tiger skin robes in Tibet, driven once again by increasing prosperity. With few tigers elsewhere, India became the obvious target.

Project Tiger had won for the tiger in India a tenuous prosperity, but unfortunately this overlaid a simmering resentment amongst the people who were displaced to assure the tiger its future and who suffered the depredations of the tiger and its prey with little or no compensation. Such people provided ready recruits to the poaching syndicates which now pose a genuine and potent threat to the very survival of the tiger. Illicit trade distorts all relationships. In the past the caste of hereditary skinners, who inhabit the fringes of every Indian village, would race to a tiger or leopard kill in advance of a vengeful owner, to harvest the skin in as intact a form as possible while usurping the kill for themselves. This curbed retaliatory poisoning — an intended consequence since killing cows is unacceptable in India and skinners have to depend on natural deaths. The tiger was an ally in this situation. Today this same caste participates in the killing of and harvesting of tiger skins and parts. Tigers are worth more dead than alive.

Fecund creatures that they are, tigers can absorb a high degree of mortality — perhaps up to 20 per cent of the total population annually — with no long-term ill effects on the stable breeding population. But so great is the current demand that the likely depletion of the tiger population is reaching beyond the 'doomed surplus' of the forest edge to the settled breeders of the core. Struggling to survive in a landscape overrun by impoverished humans and their cattle, and under constant threat from the poacher's gun, the tiger also has to contend with the pressures of rapid economic development across the countries of its range. Massive infrastructure projects are tearing apart what is left of its world; roads, dams, mines are destroying and fragmenting habitat at an unprecedented rate, leaving the fate of this charismatic predator more uncertain than ever.

Thursday November 2, 2006

Yesterday there was no sign of the mother tiger. We filmed the cubs alone for a short while, before they disappeared into the thick lantana. Today was another slow morning, which gradually became filled with disappointment, as we couldn't find any tigers. It looks like the cubs are hiding, as there is still no sign of either their mother or the male she may be leading away. Perhaps she is a long way away. It was sunny after the cloud of dawn, but the pink sky foretold of more cloud to come. It still seemed like paradise as we came out of the forest of shafted sunbeams into the clearing of the edge of the lake. It soon became warm, with a stillness amplified and broken at the same time with the calls of bulbuls. It was as if time stood still and could last forever. But we returned to the forest, and then the cloud came and a cool breeze changed everything.

Every cloud has a silver lining, and today's certainly did – cooler conditions meant that animals were still active a little longer into the morning that usual. Shivpujan found a Sloth Bear still out, as we searched through a thick patch of young teak trees. These trees have huge leaves so there isn't much visibility in these mini forests of saplings, but we glimpsed a bear rooting around like a lost shadow with its pitch-black fur. It carried on relatively undisturbed as we approached. It let us follow it for about 20 minutes as it gradually made its way up the hillside to the caves of Kalaphad, where the terrain got too steep for the elephant to follow. It gave us a nice little sequence of a bear minding its own business, digging at roots in the forest.

If it hadn't been so steep, or if we had found the bear earlier, we would certainly have had longer with it. But then any bear material is rare, as they are mainly active at night.

A Sloth Bear cub rides 'piggyback' style on its mother's back; this behaviour is unique to this species among bears.

A baby langur monkey investigates the remote camera after catching its reflection in the lens.

Today was a remarkable day as we saw the tigers from our car even before we got onto the elephants. Suddenly it felt so easy, after a couple of days of not knowing where they were. They seemed to do everything this morning, which meant we got a lot of film. In fact we definitely got the best material of the trip so far, all of it valuable for the story.

They went for a long walk, all five of them, and today we were able to film them all together, as the mother would stop, and then each cub would come and rub against her head briefly and then lie down beside her, before they moved on again. It was as if they were pleased to be back together again,

We even filmed a hunt, which was incredible, as we are now beginning to get a feel of how she does it. She works mainly with the element of surprise, seeing the prey before it sees her. She is very quick to spot prey when she is actively looking, which she was this morning.

There is a high density of deer in Pench. The forest structure and distribution of the water dams is ideal for them and they flourish. Therefore it's also ideal for tigers – they are never far from a potential meal.

Three Sambar were moving by a *nyala* when she saw them, and she went carefully for one straightaway. It hadn't seen her at all, and the attack happened so fast it was impossible to keep track of the tiger – she managed to vanish for a moment before she made her bounding leaps straight onto the Sambar. All I could do was to film the deer as best I could from the moving elephant, knowing the tiger was close, and in the

The tiger's markings and colouration allow it to blend in perfectly in the afternoon light.

blink of an eye she was on the deer. It all seemed to happen behind leaves, which always appear in front of the lens when something exciting is happening! Only by replaying the film later could I see where she had come from, even though I thought I knew exactly where she was. Geoff had also managed to catch something of the final leap on Tuskcam but it was over in a flash – the poor Sambar had no chance.

The cubs quickly appeared, as they were not far behind the mother, and immediately they started to jump on the carcass and grab it around the mouth, just like they were killing it themselves. They played like this for a while, just like big kittens playing with a large toy, all classic cat play and great for the film which I really feel is coming on well now. This kind of play is one of the vital parts of the development of young tigers and I was very pleased to have been able to film it. John too was delighted, when I made the daily phone call to him back in the UK.

A young female peers right towards a remote camera!

Wednesday November 29, 2006

We have no Mohan today. He is in serious 'musth' so may not be available for the rest of the trip. All male elephants experience this explosion of hormones about once a year. He has secretions streaming from his temples and a constant dribble from his penis, which release a smell to attract any passing receptive females. Males become very unpredictable and therefore dangerous when in musth, so he has to be restrained as a precaution, and he's absolutely not allowed to work. It's just one of those things we have to accept on a long-term project like this. A shame for us, but thank goodness we still have an alternative with Jung Bhadhur.

The season is moving on. My tea now arrives at my tent at 4.45am – the days shorten as winter approaches. The sun is noticeably lower in the sky and it's dark until 6.10am. We go into the park as soon as we can.

It seems incredible in this day and age to have a job that involves travelling on an elephant looking for tigers. There is so much else going on in the world outside. I marvel as we break out of the forest to the still, mirrored waters of the lake which stretch into the misty distance, a scene of complete tranquillity. We stop and listen, and hear only bird song and the rumbling stomach of the elephant. The sun pushes up from behind the trees, with weak rays in the mist that the lake creates. As the sun climbs the magic of the early morning anticipation reduces, until by about 9.30am there is a growing sense that we are not going to find tigers on this perfect morning.

But then we do! We find them in the open, with a fresh dead Sambar, which they must have caught as it came down to drink. The cubs were all over the carcass, running and jumping at it as we arrived, again pretending that they were killing it themselves. The mother though was clearly committed to getting it away from the tree-less expanse of the open lake, to shade on the other side. She hauled the Sambar, which was larger than herself, into the water, and then using its natural buoyancy she swam with it, floating it beside her until she reached the other side.

She left it in the water for a while, and the cubs got very frisky as they played around the carcass, they then tried to eat a little of the rump. It must have been the annoyance of the crows calling overhead that finally triggered the tigress to move it out of the water, which she then did with the help of the cubs. It was a very steep bank the other side, with a fallen tree to clamber over but that didn't deter any of them, though needless to say, the effort puffed her out. Her determination to get the deer under cover was extraordinary – eventually she succeeded and then they all collapsed into the thick shade of the chosen bushes.

It's impossible for me to go all morning without a pee, as I have to drink a lot in the heat, to keep a clear head. This presents a problem when spending the day on elephant-back, as the safest and easiest place to pee is by a high rock, so I can get off and back on easily. Shivpujan has taught me to climb down the trunk of the elephant for those occasions when there is no rock available. It is exciting suddenly being on the ground – it heightens the senses, especially as climbing back up is a lot more demanding, and may of course need to be done in a hurry… Today I found myself standing on a rock by the elephant having a pee when a deer came running towards me, hotly pursued by a big tiger cub having a go! Unbelievable!

I froze behind the elephant and luckily the tiger bounded by, oblivious of me as it chased the deer. It's an image that will live with me for a long time, not least because I wasn't able to film it, but also because it was so dramatic, and exactly what we are trying to film but never see! It is so typical of the experience of wildlife filming. What you miss you live with forever. Nobody thinks to ask what you missed, but in these conditions I miss a lot!

OPPOSITE: *The Tigers project allowed Mike to film the tigers from some amazing and unique perspectives!*

I t was an interesting day, as we found the tigers on the move from the car. Having the walkie-talkies is key in situations like this, as it meant we could immediately tell the mahouts where we were. The tigers were in hunting mode, all keen. The cubs hung back as if their mother had told them to.

We had another tense hunt, similar to the other day when the deer became confused as they realized there was more than one tiger in the area. The vegetation is against us seeing anything very clearly, but just being there is a stimulating experience. The tigress killed a deer which was close to John and Geoff and not so far from me, but my view was again behind leaves, theirs a bit clearer. It really does happen extremely fast. All the images we are getting at times like this have so much energy attached to them, you can almost feel the tension and excitement from looking at them, and especially listening to the calls of the alarmed deer and monkeys. The monkeys are very good at spotting a tiger or leopard, and when they do they will call for long periods from the trees.

The two female cubs, now almost fully grown, about to play-fight with each other.

The moment of a mock fight, an aspect of behaviour which lasts only a few seconds. They do this often as they were growing up, but each time it becomes slightly more serious and more ritualised until they will eventually fight properly and then have to separate for good to pursue independent lives.

The monkeys and the deer protect each other with their calls. They are very often together, as the deer will stand underneath the trees waiting for falling leaves from the monkeys feeding above.

After the kill we had a good show of the cubs surrounding the carcass of the deer — John grabbed an amazing still of two of the cubs play-fighting. They were hungry today, so we got more film of tigers eating. They now take it in turns. They are so big now and are aggressive to one another in more and more situations, especially where food is involved, but the actual business of eating is a bit of an anticlimax, as they appear to be so orderly. All too soon it's over and we are on our way back.

John and Geoff worked away at Trunkcam and Tuskcam to perfect their designs. There are always refinements, constant development and servicing. It always amuses me to see Geoff's room, a bed stranded in the middle of some mad professor's workshop.

A scene that hasn't changed for hundreds of years – the helpers walk the elephants back to their camp at the end of the day.

Later, I went back to the park to meet the elephants. I filmed them silhouetted out in the lake, almost submerged with the tiny figures of the mahouts' helpers balancing on them as the scrubbed their skin, in a constant movement of splashing water against the setting sun. The expanse of the lake reaching into the horizon, and the black line of the mysterious forest etched like a woodcut onto it, produced an image that looked like the India of our imaginations – the vast India waiting to be discovered. As I film them walking along the dam wall, cutting across the orb of the sun, I feel overwhelmed with emotion, and I wonder if these scenes, this place, could possibly last another millennium.

Just as it must have always been, it is the job of the mahouts' helpers to look after the elephants when they are not being ridden. These elephants have a routine, which they are very used to. Their bath is followed by a treat, in the form of giant chapattis, which the helpers cook over charcoal every morning. Sometimes they have a bowl of boiled chickpeas as well and, depending on the season, they may be given sugar cane instead. They really look forward to this. It is a joy to watch them walking away with their chapattis held by the tip of their trunks against their mouths, as they eat one at a time, walking off into the forest for the night.

THE COST OF TIGER CONSERVATION

The majority of wild tigers live in overcrowded and impoverished landscapes, conditions which are incompatible with the long-term well-being of the species. To achieve their survival in areas densely honeycombed by human habitation entails a considerable human cost. People have to abandon ancestral homes, villages have to be re-settled, lives re-built, social and economic networks re-established — and the political costs calculated accordingly. An added complication is the scarcity of suitable alternative land, a situation requiring hard choices by wildlife managers as to which area to consolidate by removing people and which to sacrifice by re-settling them there.

While successful and sympathetic resettlement and rehabilitation can be achieved, it is only part of the story. Strict wildlife reserves lock people out of what is effectively their 'supermarket', their traditional source of fodder, food, commercial products, timber and fuel. In a subsistence economy this is a massive loss, and one compounded by the fact that the restriction is only one-way: tigers feeding on livestock (and occasionally humans), and their prey feeding on crops, continue to raid the herds and fields in increasing force as their numbers multiply in the favourable environment of the reserve.

Depending on their proximity to park boundaries, it is estimated that farmers across much of India's 'tiger country' lose some 10–20 per cent of their crop annually to wildlife. There is no crop insurance or compensation system in operation, although a rudimentary scheme does operate for livestock killed by predators — but only in areas open to grazing, primarily because the connection between cattle kills and the retaliatory killing of tigers is well understood. This anomaly is typical of a conservation policy framed by an urban elite insensitive to the quotidian anxieties of life at the edge of the jungle. But the fact of the matter is that loss of crops, livestock and access to forest resources are all major economic and political issues, and any reckoning of the cost of conservation that does not take these into account will be flawed from the outset and simply store trouble for the future. In a sense this has already occurred in India, where simmering resentments over the whole issue of forest rights for the traditional inhabitants have manifested themselves in the Tribal Forest Rights Bill — a piece of legislation with the potential to completely disrupt the country's entire wildlife and habitat conservation apparatus.

The problem is aggravated by other issues, not least the fact that the poorest on the land are as much under threat from a *Blitzkrieg* of pressures as are tigers. Every large infrastructure project — be it a dam, mine or road — displaces humans as inexorably as it does tigers, and in the absence of effective relocation and rehabilitation

A tiger pug-mark in soft sand. A great deal of information can be gleaned from these simple footprints.

Tigers have to cross roads as part of their daily lives, especially where they are forced to live close to humans.

measures these people become illegal squatters in forest lands, only to be moved from there to make way for tigers! It is clear that in India the biggest failure in this respect is by the plethora of government departments charged with the task of rural development. If these were only more competent and effective they could make a sizeable difference to the rural economy and at least help alleviate some of the pressure on the forests.

This is not an attempt to paint a doom and gloom scenario but rather to highlight the complex issues that underlie the realities of tiger conservation. It is easy to forget that tigers in India and Nepal live amongst human population densities equal to those in England and western Europe. The situation in south-east Asia may be somewhat better in terms of human pressure per square metre, but is just as complicated in overall conservation terms because these societies lack the head start that India has in terms of conservation. Nor do they have, to quite the same degree, the innate religio-cultural celebration of all life that still exists in India and continues to help protect wildlife to some extent.

The human dimension to the cost of conservation is of course being recognized and innovative solutions are being trialled. The Terai Arc Landscape project between India and Nepal, where

villages at the edge of the Chitwan National Park receive a 50 per cent share of tourism revenue in exchange for protection of the buffer zone forests, has been a success story. As has a six-year effort there to reafforest and revive forest corridors between disjunct core areas under the stewardship of forest-user groups, who gain a share of forest resources while simultaneously improving the dispersal potential of tigers and other wildlife.

Although it is currently concentrated in only a handful of tiger parks with a reputation for visible tigers, tourism has tremendous potential to alleviate stress in forest-edge villages. At the same time this potential should not be equated with that of Africa, where human populations are vastly lower than in Asia and land generally less productive agriculturally. In Sumatra, for example, no amount of tourism will earn local farmers and government the same revenues as palm oil cultivation. Nevertheless, a mix of imaginative solutions is available – for example, carbon credit trading can earn income while preserving vital forests, and with all the money available in the conservation world one very viable alternative is to make direct payments for the protection of key species.

Thursday January 18, 2007

This winter's break was shorter than usual, but the season has moved on and we are now able to recognize the changes, as this is our third winter. It's very dry now and the trees give a real feeling of autumn. Everywhere is yellow with dead and dying grass. It's dusty on the road and clear in the sky, still as ever, with more silence now, in contrast to the clatter of excitement in the weeks after the monsoon at the end of last year.

Tiger cub yawning.

Out at dawn, we eventually got past all the paperwork and into the park. It was cold, very cold, as we drive in the open-topped car, the wind chill making the air feel sub-zero. It wasn't of course, but you could be forgiven for thinking so. All the staff at the gate have towels wrapped round their heads, and their bodies are wrapped in blankets, standing round little fires of dead leaves.

Sad news: the baby elephant Pench Bhadhur, Sarwasti's son, has died at only two years old, from 'swellings in the head' apparently. It is really surprising as he had such spirit. The consensus is that it is likely to be a genetic thing, as Sarwasti had another baby in the past that died in the same way at the same age.

Sarwasti was not working, as she is in mourning for her baby, so we went out on only two elephants, the two boys. Mohan's musth is over and he is quite his old self. He always amazes us with his reliability; he had no trouble remembering again how to pick up the camera, even though he has been off for a long time.

We found nothing for ages. It was quite eerie. Then suddenly there was a tiger cub by the road. The rest were not far away. They were by a dead Wild Boar, which they didn't seem very hungry for. I was a little surprised as they are hard to catch, and wild pigs are an absolute favourite of lions in Africa. It was great to see the cubs again, looking so big now. They stayed in the area and tried to conceal the carcass, the mother pawing the grass to try and cover it, to keep it from the eyes of vultures and crows.

OPPOSITE: *It took my breath away to find that all the cubs looked like adult tigers.*

Yesterday we saw no tigers, but did obtain some beautiful film of an adult male leopard, who allowed us to follow him for an hour as he went about his business — a great opportunity to observe this usually shy cat.

Today we were with all the tigers all morning, as they had two carcasses. It was difficult to surmise what might have happened. Both kills were Chital, one a large fawn which might possibly have been killed by one of the cubs, and the other an adult, perhaps the fawn's mother, which was presumably killed by the tigress. It must have happened a while ago, as both carcasses were mostly eaten — with five adult-sized tigers to feed, one carcass doesn't go as far as it used to!

It was a good situation for the remote cameras, and one of our newer developments is another camera for Mohan to deploy. This is a remote camera that looks like a log, and works in the same way as the other remotes that have been the mainstay of so much of the film to date. John has named it simply 'Logcam'. It is another great device, as it is so simple to use. Keyshu, riding Mohan, passes the camera to his elephant's trunk, and Mohan lowers it ever so gently onto the ground at the command of Keyshu. It's wonderful as we can put it down and retrieve it so quickly that we are getting shots that simply wouldn't be possible any other way.

Logcam proved to be a great new device that was easy to deploy, and Mohan seemed to get the hang of it very quickly. He was always incredibly careful and never dropped it. Geoff had made it with a handle on the top so that he could easily get his trunk through.

I have enjoyed filming him doing it, as it really does show the camera being immersed into the tiger's world, while the tigers carry on as normal.

It was a good morning; between the three of us we took great material of the cubs playing a bit, and generally being themselves. Geoff had Trunkcam today, as a carcass is an ideal situation for it, while John was having a great time taking stills and filming at the same time!

We enjoyed a great afternoon session too, as we found the tigers from the car. I had returned to the lake area and spent a long time not hearing anything. Then, just as it started to get dark and we had to leave, there they were, two tiger cubs marching out of cover and looking at the deer. There were a lot of deer so they spotted the tigers quickly, and as we were very close the chorus of calls seemed incredibly loud – I could understand how it can carry such a long way.

The cubs went straight into hunting mode as soon as they saw the Chital, and it gave me a great piece of stalking with the deer in the same shot. This kind of shot is never easy to get. For a moment I thought the cubs were going to surprise themselves – and us as well – by catching something. As the deer stood and stared as if they couldn't quite believe that a tiger was coming straight towards them. Then they turned and went bouncing away, with their white-tufted tails fully raised, vanishing into the undergrowth, leaving the cubs disappointed this time.

We were often within spitting distance of the tigers, by now we were very much part of their world.

Yesterday I somehow managed to fall off the *howdah* and onto the car roof while dismounting at lunchtime. I strained my arm badly – it was extremely painful and soon became quite swollen, which put paid to the afternoon's filming. The doctor gave me some pills for it and, to my relief, it felt much better today.

The cubs relax on a rock and wait for a deer to wander by. Deer would often appear at random, but the cubs over-exuberance would result in failure to catch their prey.

We were with the tigers from first thing today. The cubs now are hunting for themselves, walking as a group through the forest, and chasing any deer they could get anywhere near. It was a useful morning for us as there were many chances for the running cubs, but they didn't get very close to catching anything.

The mother tiger was hunting as well, and again it looked like she was using the inexperience of the cubs to her advantage. The deer fix their attention on the tiger they can actually see, and don't notice that another is creeping up on them. She caught a deer in partial vision to us, a real excitement, as it's such a charged atmosphere. The cubs were hungry today, and it was reassuring to see all four of them still eating as a family.

OVERLEAF ABOVE: *The tigers would often select the top of a rock as a resting place as it offers an obvious viewpoint; something that can be difficult to find in a forest. This cub is showing its frustration at seeing a monkey out of reach in a nearby tree.*

OVERLEAF BELOW: *The cubs on the move looking for deer or anything else they might be able to catch. It is a remarkable sight to be able to watch four near-adult tigers move determinedly through the forest.*

John managed to lose his stills camera, attempting some extremely difficult shots while tracking with the tiger during one of the hunts. After a lot of searching the camera was finally spotted – in one of the cub's mouths. The cub was walking around with it as if it were a trophy. We managed to recover it eventually and, unbelievably, it still worked.

A Sambar carcass is devoured by two of the cubs.

Wednesday February 7, 2007

One of the female cubs after a meal. She will probably stay there for the rest of the day. Sometimes they would remain even longer after a really big feast.

This morning we found the tigers again for the tenth day running. It had been wonderful to be able to be with them so much. They seem to be staying within the core area of their range — the safest option for them I think. The cubs obviously know the territory really well, much better than us; they are regularly left alone now. They will be ready to go off on their own in a few months, and the mother has to encourage them to do this. It's all very good material for our story.

We did a lot of walking this morning, and the tigress saw a Sambar female with its fawn on the edge of the lantana bushes. With the advantage of being in the forest, the tigress was able to surprise them and catch the fawn quite easily, as it was still quite small.

Sambar are always in smaller groups than the Chital, so have to be very alert, relying on their huge ears and keen vision to detect danger. They spend a lot of time standing very still, which I think must enable them to assess any threat as it appears.

The mother gave the fawn to one of the cubs to play with while it was still alive, to give them the practice of killing live prey. Not nice to witness, but it's an important piece of cat behaviour and the first time we have been able to film it with this family.

A Sambar deer nervously raises its tail; this is classic alert posture.

MYTHICAL TIGER – *Fire in the Jungle*

On a winter's afternoon in central India, as the shadows lengthened and pooled in the hollows of the hills, a man died in a simple hut built just on the boundary of a small farm at the very edge of the jungle. His name was Gulab Singh and he was a Gond, a member of a proud and ancient people who had once dominated the rugged heart of India. He was also a healer of the old type, respected – almost revered – as much for his sure healing touch as for the integrity that refused even the tiniest payment as being offensive to the Gods whose instrument he believed he was. Even the egalitarian Gond called him *Thakur*, meaning lord or chief. They said of Gulab that he could sense the future and that when he fell ill he knew that his time had come.

Refusing treatment, he was determined only to survive until the master of the house of which he was caretaker returned, releasing Gulab Singh from his responsibility. But that fierce spirit ebbed before the owners' return. His sons were with him at that moment and, since custom decreed an immediate cremation, they stepped out of the house to return to the village and gather everyone for the last rites. In so doing they came face to face with the most enormous male tiger, lying just a few feet from the entrance to the hut.

As the shocked brothers leaped back inside, the tiger moved off calmly into a bamboo clump, allowing them the space to escape. This they did at high speed, convinced that the tiger had come to lift the dead body and that their own lives were at stake. Reinforcements gathered, they returned with much banging of pots and pans but the tiger had disappeared and Gulab Singh lay undisturbed. After the cremation, men gathered round the fires and spoke deep into the night of tigers and the spirit that presided over that jungle, the *Sidh Baba* (the guardian spirit) with whom, as was well-known, Gulab Singh had communed and to whose will all tigers were subject. The next morning, when the brothers returned to the hut, the tiger was there once again. By the time the owner returned, everyone in the district knew that although Gulab Singh's body had failed him, his spirit had summoned a tiger which was now guarding the house until the owner spent a night there, thus releasing Gulab's spirit from its self-imposed duty.

This remarkable and true story vividly illustrates the central place occupied by this magnificent cat in the mythic landscape of the human cultures with which it shares a home. Stories like this abound across 'tigerland', for this is a creature that touches some primal chord in humans, going far beyond the mere fact of its physical beauty. Elusive, powerful, almost a phantom presence, the tiger embodies all the majesty and mystery of the wilderness over which it presides. The Russian biologist Vladimir Troinin quotes the old hunter Udegei, "The Taiga has a great soul. The tiger is part of that soul…..If you kill a tiger, the Taiga is weakened. If you kill all the tigers, the Taiga will lose its soul entirely."

One thing is certain: if the tiger were ever to vanish, it would leave the forests bereft of colour and drained of that vibrant energy, that magic sparkle that is the gift of the great predators. Too often in this modern world we seem constrained to deny the honest emotional response to the great creatures of the Earth that make our senses leap and fill our beings with exhilaration, as an adequate reason for preserving them.

Everything seems to hinge on economic arguments and we seem constantly to be pressed into corners, reduced to justifying the survival of a species with the logic of the bazaar.

When the Indian Prime Minister Indira Gandhi threw her considerable political weight and matchless prestige in the early 1970s behind the effort to save the tiger by initiating the world's most successful conservation programme (at the time, at least) in the shape of Project Tiger, she was moved wholly by the urgent need to save this magnificent species. Her decision was untainted by commercial considerations, but this is not to deny economic realities. We have already seen that conservation carries costs and that no effective long-term conservation is possible unless these are taken into account. Equally, conservation bequeaths undoubted benefits and any activity that can defray the costs of conservation while preserving biodiversity must surely be adopted – be it carbon credit trading or responsible tourism. However, we must never permit the economic argument to be paramount unless we can reconcile ourselves to allowing the fate of the tiger to be governed by the vagaries of the market place. Surely the 'tiger' economies of the tiger-range countries can afford to set aside the modest sums and investment in political will that are required to preserve the grand creature they invoke as a measure of their own vitality?

Tigers feature heavily in many Asian cultures.

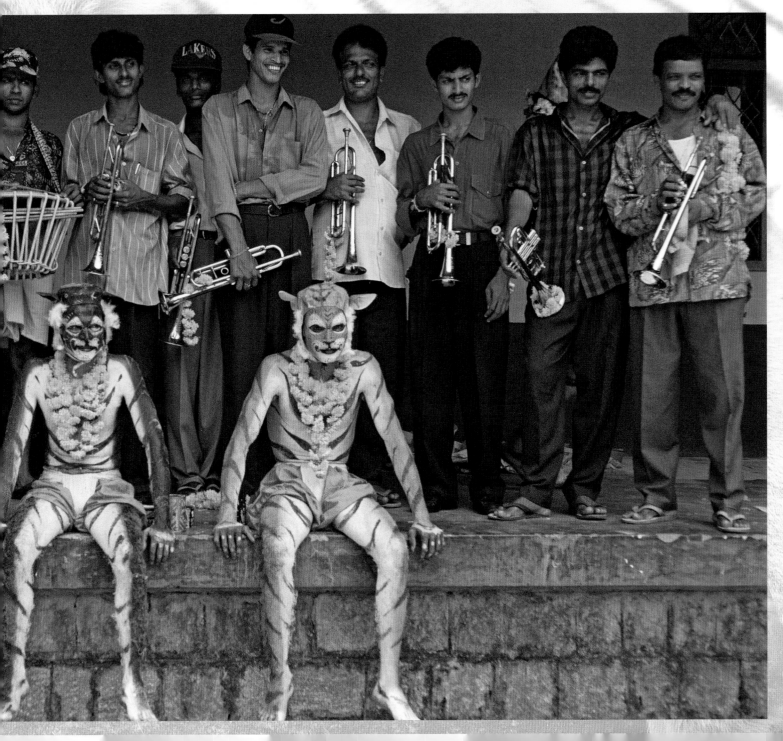

Friday March 23, 2007

We were with the tigers today as usual; we seem to find them all the time now. Yesterday we had filmed a female cub practising hunting by herself, and then later one of the males, also on his own. Today the cubs were very loosely together, so it was impossible to be able to follow them all as a group.

The cubs are separating themselves naturally. The two males will probably stay together and be best mates, surviving in or near the edge of the territory for as long as they can, usually until their father pushes them out for good. They will then be out on their own in an unfamiliar place, where they have to find females of their own, fighting other males for the opportunity to mate.

It will be tough for them. One of these cubs is very fond of his mother. She still lets him rest his head on her back in the rare moments when they are all together. He will miss her company.

One of the cubs rests in a favourite place which benefits from the breeze and the shade of bamboo. They seem to look right through you as if you weren't there.

The two female cubs will tolerate the males but are becoming very aggressive towards each other and will not share anything any more. They will have to separate in a few months, and set up a territory of their own. These are likely to be within their present home range, as it's so very large. Their mother should tolerate them in different parts of it.

Today we watched a male and a female cub together. I was wondering how they'd survive, as they do not appear to have the finesse of hunting that their mother has. I had an answer very quickly, as the male suddenly had a partly eaten langur monkey in his mouth, which was to keep them in one place for the rest of the morning.

The monkey was the remains of a leopard kill, fresh, but with no sign of the leopard. The leopard will only eat what it can, before leaving it. It's another surprise to me, as the leopards here don't take their kills up into the trees as the leopards in Africa do. In all the time we have been filming here we have seen leopards up trees, but never with food. It was a nice bonus for the tiger, as it involved no effort on his part. Even so, although he looked like he wanted it all, he did share it with his sister.

We only got half a day's footage yesterday, as there was more trouble with fish poachers and the elephants were not available until lunchtime. We did film the cubs in the afternoon, capturing interesting behaviour including the two males play mating, and then one female stalking another – this looked like play, but it ended very aggressively.

Today it really looked like we were not going to find any tigers. It became so hot, as it always does now, and any hope tends to die away after about 9.30am. But at 9.45am, we did find one female cub desperately trying getting cool under rocks away from the sun. This was not far from the lake, a place where they have been before – 'tiger house', as the mahouts call it.

There were signs that they had made a kill out on the lake earlier, Jung Bhadhur had visited there but had found only pugmarks, with no real clues as to where any of the other cubs might have gone.

In the afternoon we found a freshly killed sambar female right by the road. The thick lantana all around made me a feel a little nervous as I got out of the vehicle. But, moving cautiously, I was able to place two remote cameras by it, and to feel that the body was still warm. We were all excited, as it certainly does quicken your heart rate when it's obvious that a tiger must be nearby!

There was nothing else we could do except leave the cameras and, for the rest of the afternoon, watch from as far away as we reasonably could without losing sight of the carcass. Disappointingly nothing came to the kill before sunset, but we left the cameras in place, feeling that they are bound to see something.

OPPOSITE: *Three cubs in the shade of a rock. It gets very hot and they obviously feel uncomfortable so deep shade is at a premium and the rocks provide it. Sometimes there is only enough space for one or two tigers so they have to separate, In this case there was enough for all of them, which seems to be preferable even if it means almost lying on each other.*

BELOW: *Female cub resting in the mid-afternoon heat.*

Thursday May 10, 2007

t is very hot and almost humid, feeling quite tropical now. Each morning we wake up from restless sleep wet with sweat. We got into the reserve a bit earlier today, and drove straight to where we told tigers had been seen last night. Two were still there, the mother tiger and one female cub, but they ambled away down the road before the elephants arrived. When the three elephants did join us 20 minutes later, we searched the area but the tigers were nowhere to be seen or heard.

The Range Officer (one of the directors of the park) has asked us to consider using our remote cameras in an area of the river where there is another family of tigers with much smaller cubs. There are several other tigers in Pench, living outside the area we have been working in – because they are basically out of range of the elephant camp, the mahouts rarely visit them. Several of these tigers are known to the park, as some are offspring of the more 'local' tigers, and they are occasionally seen at the edges of the ranges that the mahouts come to in their searches. This certainly explains the random pugmarks that we often seem to find.

One of the guards had seen these new cubs and he took us to the site and helped us place the cameras. He was genuinely interested in and very enthusiastic about the tigers. The site he took us to is beautiful, with big trees bordering the riverbanks. I felt the eyes of the mother tiger on us all the time, but knew she would be too frightened to come anywhere near four humans.

We placed two cameras along an obvious route across what is still quite a big body of water, and then left – we're happy for them to stay at this exciting site for several days.

We had a cup of tea with the Range Officer when we took him back, sitting under a banyan tree by the Parks radio station. He told us of a 3 m python that he'd watched as it lay in a pool catching small monkeys and birds; we have to have a go at that!

After the tea we drove back to film Sarwasti's new baby elephant, born recently (a surprise to us and the mahouts as no-one realized she was pregnant). The baby was enchanting, wading in the water and playing with its trunk. It made a lovely piece of film.

OPPOSITE: *The new calf would never stray from Sarwasti's side.*

RIGHT: *Sarwasti and her new baby, which was a big surprise to everyone.*

138

Friday May 11, 2007

Today was a real dry season day – clear and cloudless with warm air, even first thing in the morning. We found the tigers back at the waterhole where they were yesterday morning. They must come and spend the night in or by the water. We were there quite a bit before the elephants, but they arrived in time this morning.

Geoff set off on Mohan with Trunkcam. I rode on Domini, who is the smallest and fastest elephant, and we were able to keep up with the tigers.

The four cubs still stay quite close together, but the mother tiger snarls quite a bit at them now, as if she is breaking the bond between them. It's more great story material for the film.

She left in a different direction, and climbed away up to the hills in the west, but the cubs all wanted to go to Kalaphad, in the opposite direction, and determinedly did so. We got only small bits of film, as they sat by the water and then walked away.

Tiger in seventh heaven – it almost looked enjoyable enough to do the same. What is surprising is that the mud dries fast and falls off their coats quite quickly in the hot sun.

We tried to follow them up the lava rocks but this was too steep for Mohan, who was carrying the camera in his trunk. He tried ever so carefully, using his trunk as a fifth leg in the steepest bits, but it was too much. Keyshu will never force him to make an error, so he went back.

This was just as well, as for the first time for months the elephants found Charger, the male who we now believe is the father of the cubs. He was sitting in a still muddy pool below the dam, not so far from where they all were this morning when we arrived. Geoff got some material of him with Trunkcam, leaving the mud and walking away. It was a little slow as mornings go, but the heat is the controlling factor for them now, and by 9am they just want to be hidden for the rest of the day.

We went in the car to the waterhole where the python the Range Officer told us about is said to live, and put a remote camera down, but it's a tricky site, so I will not hold my breath. There is very little water and a large area of deep rocks for the snake to live in. There are plenty of bees drinking at the dwindling pool.

'Charger' in full roar. He was so named by the mahouts as he was always aggressive towards the elephants when he first appeared on the scene. The deep roar of a tiger is a frightening experience even if you are comparatively safe on top of an elephant.

otwithstanding the regular doomsday predictions carried by the media, few responsible scientists expect the tiger to become extinct in the immediate future. Despite a massive loss of habitat across the tiger range countries – only an estimated seven per cent of tiger habitat survives today, compared to a hundred years ago – a recent survey by the Wildlife Conservation Society and the World Wide Fund for Nature estimates that about 1.5 million sq km (579,000 sq miles) of potential tiger habitat still survives. Only a tiny fraction of these forests, totalling perhaps one per cent of the entire land mass of the region, presently support tigers in any viable numbers. Yet the strong potential of these forests to sustain significant tiger populations in the future is reflected in the WCS's stated goal "to have 100,000 tigers in the wild by 2101".

Tiger ecologists and conservationists have now embarked on a two-pronged strategy for the long-term conservation of the species. This is based on identifying the most promising large chunks of tiger habitat (called 'Tiger Conservation Units'), ie those areas encompassing healthy core breeding populations which can be connected through linking existing/rehabilitated forests, alongside the mobilization of a network of politicians, opinion-makers, administrators, religious leaders and human rights activists. This network will help form and influence pro-conservation public opinion while simultaneously advocating strong and effective protection of the national parks and reserves.

This is an enormous job and one that must harness science, political advocacy and imaginative, multi-layered strategies to reconcile and co-opt local populations to the task of conservation. The key point to remember is that effective solutions and strategies must be site-specific, and address local cultural, political and economic conditions and sensibilities. There can be no common prescriptive remedy. What works in south India may not be relevant in central India, and will almost certainly be alien to south-east Asia.

A brief survey of the present conservation scenario is instructive. The Indian subcontinent continues to harbour the highest tiger populations in the world, with three significant TCUs in the shape of the Western Ghats TCU in the south, the vast central Indian forests forming the Satpuda–Maikal TCU (which possibly harbours the largest tiger population in the subcontinent) and the Terai Arc TCU, mentioned earlier (see page 121) and which is shared between India and Nepal. For the present, the subcontinent continues to hold the best potential for the future of the tiger, despite the problems of overpopulation, poverty and rapid economic development. Cultural and political

The future for all tigers is in the hands of the humans that surround them.

attitudes there have long been sensitized to the needs of conservation and there is no serious public opposition.

Nevertheless, the situation in north-east India is somewhat different to the rest of the country and bears greater resemblance

It is imperative that all our efforts concentrate on securing the future of this charismatic species.

a deep-seated intolerance of wildlife competing for resources are just some of the challenges that conservationists face here.

However, governments can be coaxed into increasing the size of protected areas, as Burma has done in the Hukaung Valley Tiger Reserve, which comprises almost 22,000 sq km (8,500 sq miles) of forest. If the hunting of tiger prey can be reduced or at least regulated, the potential of this area to support tigers is enormous. The same can be said of several areas of Thailand, southern China, Cambodia, Laos and Vietnam, where tiger densities average less than one per 100 sq km (38 sq miles) – a fraction of what they once were. In Sumatra, where lowland forests are being converted to palm oil, there is an attempt to preserve the dwindling remnants. Perhaps up to 400 tigers survive in Sumatra, but poaching pressure here remains high.

The northernmost population of tigers is of course in the Amur River basin. The Russian Far East TCU has benefited enormously from protection. In the 1940s the tiger had almost disappeared from these vast forests, shot out by hunters. Today despite the development of the taiga – and the associated developing road network, which brings with it the pressure of hunters, particularly of prey animals – Amur tigers are estimated to number between 350 and 450.

So whilst tigers are not a lost cause, their future remains uncertain. There is much well-meaning debate about the practicability of introducing captive-bred tigers into the wild but the point is that if you cannot create and maintain the conditions required to protect wild tigers, what chance do captive-bred tigers have upon release? The other controversial issue is the contention that pressure on wild tiger populations can be mitigated by farming them in captivity to supply the demand for tiger parts in Traditional Chinese Medicine (TCM). The offensive nature of this idea aside, the reality is that killing wild tigers is far cheaper than breeding them in captivity and there is, in any case, clear evidence that consumers believe wild tiger parts to be more potent. Therefore any 'legal' trade in captive tiger parts will serve to mask the illicit trade in wild tigers, and aggravate the already intolerable pressure on wild populations.

Tigers are remarkably resilient animals with a very strong reproductive capacity, so long as suitable habitat and adequate prey is available. And if a silver lining can be detected to the dark cloud that currently hangs over this iconic species, than it must be their powers of survival and renewal. However, their habitat and prey are disappearing fast and it is therefore imperative that all our efforts are focused on securing these resources and then protecting them by every means at our disposal.

to that in south-east Asia, where although vast tracts of primarily montane forests remain, they have been wiped clean of their wildlife by relentless commercial and bush-meat hunting pressure. Logging, conversion of lowland forest to palm oil plantations and

Sunday May 13, 2007

Today and yesterday we filmed only a few cameos to add to the film, as the tigers were not keen to move far today. Yesterday we filmed a few moments of serious aggression between the two female cubs — they will have to separate permanently very soon. Today we found their mother lying fat, relaxed and upside down in lantana. She wasn't going anywhere.

I therefore went to look for the rest of the family with Domini. Crossing the open lake area, we found the remote camera from yesterday surrounded by pugmarks, and facing in a different direction than it had been before. This could mean some wonderful footage (we later found that there was — a great shot of a tiger's nose and whiskers). We carried on, having picked up the camera, to No. 4 lantanas where we surprise a male cub, who is clearly finding it difficult to move, He is very fat having eaten too much.

Five full-grown tigers soon demolish every last scrap of a deer carcass. There is no action now; it must have all happened under the moonless night skies, it's the best time of the month now for the next week for hunting under complete cover of darkness, which is great for them although it makes life harder for us.

We stayed with the male cub as he became more lively, after a while moving restlessly from bush to bush, then out into the open to have a swim in the pool of water. He made the usual nervous approach, testing first with the paw, before getting in backwards. He stayed in for about ten minutes, and then climbed out to another patch of lantana on that side of the valley.

It was another hot, windy day, preceded by a hot, windy, cloudless night, the sort of consistent blow that comes before the monsoon.

OPPOSITE: *The first signs of the cubs going their own ways. The female cubs have become very aggressive to one another and now there is much more serious intent in what used to be play behaviour.*

BELOW: *The mother tigress sleeping alone in the forest.*

Wednesday May 16, 2007

Two days ago we found one of the large female cubs very much on her own. We were following a set of pugmarks down the track. She was by a small dam, much further down the river. She looked fat and sleepy, as if she had been eating at night, although there was no visible carcass in the open area she had selected. This looked like a possible end to our story, as it shows one of the cubs now surviving in an area, which is within but on the edge of the territory, therefore tolerated by the mother.

Today was a quiet day. By noon we were waiting in the shade with the elephants after our fruitless search for tigers this morning, eating a late breakfast of bananas. Mohan was watching me so I gave him the skin, which he loves of course. Domini was watching too, so I gave my second skin to her. They don't miss much.

The two of them were standing side by side. They kicked the dust in small piles, then picked it up in their trunks and blew it on their stomachs, anything to try and relieve the heat. It was really far too hot for them and us, so we had to give up.

It had been a disappointing morning as there were no signs, no tigers, and therefore no film today.

We had just the two elephants today, and it looks like this is all we will have every day from now on, as Jung Bhadhur has boils on his back, and has a doctor's note to be off for a while. The vet will decide when he returns to work, but it looks like we will have to finish the film without him. They treat the elephants very well, and the park vet, who has a car and medicines and the responsibility for their health, watches over any sort of injury or ailment. Sarwasti, of course, is still on maternity leave and will be for many more months while she cares for her new baby.

OPPOSITE: *The elephants are constantly eating and seem to derive nourishment from virtually any vegetation they can get hold of.*

BELOW: *Jung Bhadhur moves silently and swiftly through the forest. A boil can be seen clearly on his back.*

Not quite ready to face the world, this male cub still has a bit more growing up to do.

loved this morning, not because it turned out to be our last but because it was still so full of expectation right from the start. I didn't want to believe that it was all about to end. Although we have not had much luck with the tigers lately, today it was as if the mahouts had really known where they had been all the

time. We found them, all of them, just as the temperature was rising and I was beginning to face the realisation that we might find nothing at all.

It was a very relaxed scene – the whole tiger family lying in the forest, centred round a recently killed Wild Boar. They all kept their distance from one another, taking very clearly defined turns to feed – one would walk away before the next roared onto the carcass to claim possession.

One of the female cubs having a scratch and looking almost like an adult now.

This was the first time in two weeks that we had seen them all together, as it looks like they are beginning to go their own ways. It could be the last time we see them like this as we only have a few days left, so I had mixed emotions watching them there, sadness that I will not see much more of them, but happiness to think of them protected here in this park, safe for now in the hands of people who I think do really care about them. At 20 months old the young tigers will still come together occasionally, their mother will still be there for them for a few months more. But by the time they are two years old they will have to split up for good, which will happen under the lush wet cloak of the monsoon. I drew in deep breaths to savour these last moments of peace in this particular place, knowing it's the end of our time here.

We returned to the site far away on the riverbank to collect the remote cameras we put down with the guard some time ago. There were pugmarks by one of them, and the film revealed a new tiger family — a

mother with four small cubs. It was a great piece of film and a lovely end to our story. It's wonderful to see a new family of tigers, and great for the park. They are living in a very special area which is stunningly beautiful and gives a real sense of hope that such places will have tigers for ever. I felt a tremendous surge of excitement watching them appear on the screen, giving such a fitting end to our filming here. We have always known there are other tigers in places we couldn't go to because they are too far away, and now seeing this family living on the river I can only feel optimistic for the future.

I know the park authorities will do everything in their power to maintain the balance of this park for the benefit of the tigers and other wildlife that lives here. The tiger is an iconic animal, and everyone wants to see it. The extraordinary thing is that a predator with the stature of this animal can be seen by lots of people. The tigers really don't mind us gazing into their world, they've grown up with it.

So, the challenge is not finding them, but controlling our access to them. With an increasingly mobile and huge human population there is a challenge here, which the park's very existence is helping to solve. Hopefully, we can go away happy that the tigers and the people of India can continue to live side by side and breathe the same air, in the peace and security of a place like Pench.

At 20 months old the cubs still meet up occasionally and walk with their mother. In another few months they will have separated and gone their own ways.

151

ACKNOWLEDGEMENTS

John Downer Productions is a UK-based production company which has a string of extraordinary films to its credit. John has always made films that push the limits of technical possibility. His creative drive and inspiration is unending. He realizes his vision with the help of his able staff, most of whom were directly involved in the tigers project over the two and half years that it took to make the films.

John had initially sent Tilly Parker from his office to evaluate the possibilities of making a film about tigers. Emma Ballinger arranged all the logistics for each trip, helped by Toby Sinclair and Suhail Gupta who encouraged us. They were able to help with all the permissions from The Ministry of External Affairs and The National Parks Office and then locally with the directors of Pench National Park itself. They dealt with all the 'red tape' and gave us a smooth passage to Pench for all the many trips we made. Toby helped with lots of permissions for film crews, and Suhail arranged our guides. Salim, Dyshuant and Aditycha were guides on our first trips. The guides were a vital link between us and all the staff of the park. They drove us around and interpreted for us, to ensure we were able to film within the park's rules. Digpal

Karmawas – or 'Diggi' as we called him – did the lion's share (so to speak) and reliably fixed everything right until the end of the project.

A big thank you to the staff of the camps Bagh Van and Pench Jungle Camp. We are indebted to the directors of the park who embraced the film and gave us all the help they could, particularly Mr Dongrial, the Field Director, Mr S. Sen, the deputy Director and Rajnish Singh, the Ranger Officer; also the Wildlife Office in Bhopal for giving us permission to film. We would also like to thank the Government of Madhya Pradesh.

The mahouts and their elephants were an indispensable team who were always there and keen to find the tigers for us, and without them there would not have been a film – Keyshu with Mohan Bhadhur, Shivpjuan with Sarwasti, Satyanarayan with Jung Bhadhur, and Sadahan with Domini. They in turn had a team of helpers whose sole job is the preparation and well-being of the elephants and we thank them too – Jainatah, Balam, Ramesh and Santosh.

INDEX

Page numbers in *italics* refer to photographs